Beazley's
Design and Detail
of the Space between
Buildings

Old and new materials of different kinds can be successfully brought together provided that the overall objective is clearly understood.

Beazley's Design and Detail of the Space between Buildings

ANGI and ALAN PINDER

Eastabrook Associates
Architects and Landscape Architects,
Stow-on-the-Wold, Gloucestershire

London
E. & F.N. SPON

First published in 1960 as The Design and Detail
of the Space between Buildings/*Beazley*
by Architectural Press Ltd
This edition published in 1990 by
E. & F.N. Spon Ltd
11 New Fetter Lane, London EC4P 4EE
Published in the USA by
Van Nostrand Reinhold
115 Fifth Avenue, New York NY 10003

© *1990 A. and A. Pinder*
Typeset in 10/12½ pt Palatino by
Photoprint, Torquay
Printed in Great Britain at the
University Press, Cambridge

ISBN 0 419 13620 7

British Library Cataloguing in Publication Data

Beazley, Elisabeth, *1923–*
 Beazley's design and detail of space between buildings.
 1. Great Britain. Landscape design.
 I. Title II. Pinder, Angi III. Pinder, Alan
 712

 ISBN 0–419–13620–7

Library of Congress Cataloging in Publication Data available

CONTENTS

INTRODUCTIONS TO THIS EDITION

This new edition, like the first, is a down-to-earth handbook so the revision has involved most exacting work. The book has also been greatly expanded. This has been entirely undertaken by Alan and Angi Pinder. It has included the checking of innumerable references – to organizations and their publications (some now defunct and replaced by new ones), to British Standards and to trade literature. The book's conversion to metric (imperial measure remains for use in the USA) was itself no mean task.

It remains a practical book telling how things are done and why, rather than what to do. As in the First edition the reasons which lie behind much of the best traditional design of the space between buildings are again explored (this was the part which was fascinating to think about and write; the rest was a hard slog). People have kindly said that it was this background which made the book particularly useful in schools of architecture and landscape design. New ideas have also been injected here and, with many new illustrations, have produced a new version of the book which over the years has made me many friends. I believe it will do the same for the Pinders. I am most grateful to them.

Elisabeth Beazley

From the date it was first published, this book has remained unique, both in its scope and in its approach to its subject. Even though the First edition has been out of print for a number of years, it continues to be referred to as an important source.

To our minds, its greatest value is that it discusses the topics it covers from the holistic position of a designer. It does not simply offer standardized technological answers to simplified technological questions. Rather, it opens each of its sections with a discussion of broader issues. From there it is able to establish the principles which, in varying weights of combination, will condition each specific design context. Only then are the potentials of different technological approaches introduced and the ways in which technology may further constrain the designer discussed. The figures are not offered as drawings to be unthinkingly replicated but as illustrations which further amplify the text.

All these qualities have been retained in this edition. References to products no longer manufactured have been omitted, and more importantly those which represent a new technological genus have been introduced. Dimensions are now given in metric, with imperial figures following in brackets. (Where precise dimensioning is not critical, these are round figure conversions.) The basic structure of the book has been changed a little. It now has three main sections. The first covers essentially horizontal elements, the second essentially vertical elements, whilst the third looks at constructional principles associated with changes of level.

There are two appendices. These allow the book to be used as an immediate reference. The first brings together, and amplifies, design and planning data given in various places in the body of the text, whilst the second supplies the addresses of the major sources of construction information.

There are three ways in which this book can be used. Student designers should read the introductions to each section, and then the main sub-sections, as a whole – concentrating on gaining an understanding of principles. With more experience, it can be used as a memory jog – finding, perhaps, the appropriate sub-section through the index. Finally, fully experienced designers will find the appendices particularly valuable as first reference that identify other, and more detailed, sources of information.

Angi and Alan Pinder

INTRODUCTION TO
THE FIRST EDITION

'A city should be built to the convenience and satisfaction of those that live in it, and to the great surprise of strangers' said San Savino in the sixteenth century. This conjures up a hundred images: Rome, Oxford, Sienna, Edinburgh, Bath . . . for 'cities' he could have laboriously said 'the buildings and the outdoor spaces created among and around them'. For, as everyone knows, it is not the buildings alone that make a town (or for that matter a village, or any group of buildings). It is the streets and the squares, the alleys and the yards, the sudden shadows in dark archways and the bright light in the courtyard beyond, the sociable openness of the market square, contrasted with the privacy of high-walled town gardens, the bustling street, and the mysterious flight of steps leading round some blind corner to end who knows where. It is the close-knit contrast that makes the drama, and the drama that surprises. One could wander for hours in the sort of town that San Savino had in mind: people do. Such towns are starred by the coach tours and extolled by the excursion organizers. Most of our best villages have realized their tourist attracting potentialities, and thousands of sight-seers gladly pay their half-crowns not only to visit the stately homes themselves, but to wander in the grounds that surround them. But who would spend an afternoon exploring Slough, or a morning doing Watford? Who would leave the village green for the council estate (although there is usually just as much grass there)? Who would wander 'round the gardens of the new municipal offices in preference to some churchyard or country house? We are too blunted to be surprised by many of our twentieth-century efforts, but not too hardened not to be depressed.

But not constantly depressed; there are some outstanding examples of good new work, outstanding because they are straightforward, robust, simple and well detailed. These virtues should be the rule, not the rare exception.

There is a strange resignation on the part of the general public, who are often oddly confused about what they like and what we can now do. Though enjoying the past tradition they do not seem to realize that a new and equally good one can, and in fact gradually is, taking its place. The present attitude towards outdoor design is one of either apathy or perverse preciousness, both of which are quite illogical. When the importance of exterior design is compared, in terms of the number of people affected, with that of the interior design of any building, this attitude is even stranger. Consider the millions who 'know' and genuinely admire Oxford but who have never been inside a college (meaning indoors), or the hundreds who walk down some street or across a square daily on the way to work, without ever entering the houses that surround them. Extraordinary care may be lavished (and quite rightly) on the design of the inside spaces themselves and the colours and textures and the durability of their materials. But the outside spaces tend to be treated either as a sort of no-man's-land where economy is the keynote, or as somewhere not good enough in itself to be designed on the same principles as the rest – a space to be prettied up in the name of amenity. There is something apologetic in these attitudes. Designers who have produced a logical building, or a serviceable highway, or a fine stretch of lawn, may become almost irrational when confronted with a piece of ground not occupied by these things. Some extremely well designed modern buildings now have cobbled forecourts which would do credit to an Edwardian tea cosy; materials are introduced, apparently with no thought for anything but variety of texture. Cotswold type walling still appears in the heart of London, looking so incongruous that one searches involuntarily for the gnomes, but it also turns up, giving a smack of inappropriate craftiness, in all parts of the country. The fake must not be allowed to blur our admiration for the real thing in its natural setting.

That is one side of the picture; the other, though equally dour, is the bleak 'serviceable' approach which takes a standard object or a standard dimension and uses it with utter inflexibility. The layout of most council estates was once an example of this; in many cases it still is. Whatever the site, the semi-detached pair has its draughty front garden, puny fence, standard path and verge, standard pale concrete kerb, and standard width of road. No wonder any zest for exploration palls as one leaves the rich diversity of village life and comes upon that. Mile upon mile of it is eating up Britain; while you read this a bulldozer is actually gnawing up some field. While our standards of building design have in many instances improved immeasurably, our standards of outdoor design have dropped to a very low ebb. To take one example: what Victorian spec. builder would have omitted the garden wall, 'round the yard of what is now called slum property, let alone, 'round the gardens of the middle-class customers? But now, chain link or chestnut pale is the rule, and this custom in town terraced houses, is almost as inconvenient as it would be to have chicken wire separating the living-rooms instead of a nine-inch party wall. Inside, there are many new comforts; built-in cupboards, central heating, television, and refrigerators. It might be thought that with all these amenities people would also claim the basic necessity of some privacy from their neighbours. In front too, emphasis has shifted. Probably as a reaction to the industrial pall covering vast areas of nineteenth-century England, grass has become synonymous with amenity. People who would not dream of putting an inexpensive felt carpet where they expect heavy wear will optimistically grass over the smallest patches or the likeliest short cuts and then are surprised or disappointed when the grass is worn brown. The robust common-sense attitude of sane tradition has changed to a sort of neurotic anxiety to make things pretty, or to ignore the problem totally.

Much of this is done for the sake of economy; labour is expensive and materials which were once taken for granted are now luxuries. Grass is initially cheaper. But there are new materials, and there are people who still feel it is worthwhile spending money on the old ones. There is also the old argument between the wisdom of facing maintenance costs, versus a higher original capital outlay. But whatever is done, should be done with an awareness of what function the detail or material really has. Paving materials are not just so many scraps of cotton for a patchwork quilt. Walls and fences have their special uses and characters, some particularly suited to one situation and utterly foreign to another. England is not a large allotment where anything goes. Street and garden furniture needs as careful choice as a dining-room table or chest of drawers.

This is everyone's responsibility. There are still (thanks be) several million private individuals who can change the face of Britain daily if they want to, without asking permission. Unfortunately this change is not always for the better. There are tens of thousands who affect the face of the land by their committee decisions on the designs of others, and thousands who consciously design our environment because they like doing it for their daily bread. With this vast army involved, there is surely a right to hope for better things.

With some very notable exceptions, the last few decades have seen the breakdown of a sane tradition, followed by haphazard indifference interspersed with bursts of grandiose pomposity. Today we have a mixture of economic stringency, prim prettiness, and misguided zeal in re-discovery of traditional materials; but there are also some very well-designed examples of the new tradition which is emerging (or old tradition re-emerging). These should be as the air we breathe.

One reason for the comparative rarity of good examples may be the scattered form which the available information on these materials and details now takes. They are no longer part of every jobbing builder's vocabulary. They exist in people's heads, in office specifications, in trade catalogues, in HMSO publications, in journals, in Victorian books on building construction, in books intended for highway engineers – the list seems endless. Much time and effort is wasted in tracking down the information. This book sets out to collect it together. It is literally a down-to-earth book about the materials, new and traditional, and details for 'hard landscape' work which is simply intended for 'convenience and satisfaction'. But it is, perhaps, from such ingredients that the things that greatly surprise strangers may also be designed.

Elisabeth Beazley, 1960

ACKNOWLEDGEMENTS

There are many people from a host of companies, trade associations and the like who have assisted with this book and to whom we extend our thanks.

Most particularly, our gratitude to John Simpson (an ex-colleague at GlosCAT) must be noted. John has not only offered general comments and encouragement but has also undertaken the arduous and tedius task of checking British Standard references, addresses and phone numbers and metric conversions.

Our gratitude to Stephanie Power must also be particularly mentioned. Steph spent many hours over a hot word-processor drafting and editing the 'manuscript' – all in addition to her regular work.

Finally, we owe a debt to Elisabeth Beazley that is quite impossible to put into words. The model offered by her first edition proved both inspiration and set a standard for our aspiration. Elisabeth has been in regular contact during the whole of the process of writing and producing this book. She most effectively resisted the undoubted temptation to dicuss every change to her original creation whilst gently reminding us of the intended nature of the whole when we became overly-immersed in detail.

Thank you John, Steph and Elizabeth.

For the illustrations, thanks are due to Aerofilms (p.xiv), Peter Pitt (p.27, top), The Architectural Press (p.62), Keith Scott (p.83, top) and Professor Pietro Porcinai (p.225).

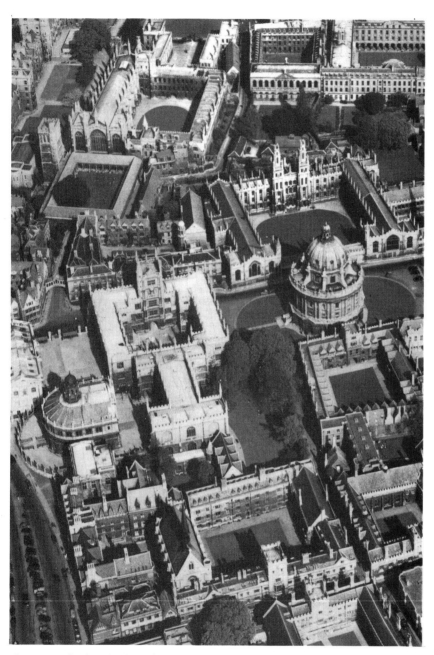

There must be hundreds of thousands who love and genuinely admire Oxford, but who have never been inside (meaning indoors) one of the colleges. This suggests that the spaces between buildings matter as much as, if not more than, the interiors of the buildings themselves.

Chapter one

PAVED
SURFACES

This chapter deals with external hard surfaces, including private roads for light traffic. Certain notes also apply to public highways. Owing to the wide range of functions and materials that are included, it is difficult to avoid repetition, unless cross-references are included. These have been kept to a minimum as it is realized how maddening they can be, and the chapter has been organized thus:

1. notes on the functions of paved surfaces
2. list of highway engineering terms
3. notes on roads and pavements primarily intended for wheeled traffic
4. general design problems common to all paving structures
5. descriptions of individual types of pavement, by type rather than function. This means that surfaces intended for vehicular or pedestrian traffic are described together. Since the problems involved are often common to both, and since many pedestrian spaces are also intended to take the occasional wheeled vehicle, this seems the clearest division to make
6. problems of detail common to all types of pavement.

1.1   THE FUNCTION OF PAVED SURFACES

The main function of any paving, to provide a hard surface, is so obvious that its more subtle but extremely important minor functions are sometimes forgotten. It is a safe rule, though it must occasionally be broken, never to change the material without a practical reason. Changing materials simply for the sake of creating pattern, can be very precious. If the traditional use of paving materials is analysed, a practical reason can usually be found for any change of material in the surface pattern. The exception to this is of course the sophisticated patterns that are consciously designed in an area of paving e.g. the lighter stone in St Mark's Square in Venice which, since it is not parallel with the walls of the square, reinforces one's awareness of the plan as a whole. But such instances are rarer than one might at first suppose. In countless traditional streets and squares, market places, courtyards, college quadrangles, formal and informal gardens and parkland, on quays and wharves, in churchyards and farmyards, change of material indicates subtle change of purpose. Functions might be listed thus:

1.1.1 Practicalities

To provide a hard, dry, non-slippery surface which will carry the load of traffic asked of it; change of traffic may indicate change of material. This leads to several less tangible functions.

1.1.2 Direction

To provide a sense of direction; paving can lead and guide people who otherwise might not know which direction to choose. Particularly good examples exist in interconnecting college quadrangles, alleys and yards. One emerges from a quadrangle into a small cobbled (say) yard; there are several doors but no apparent exit, but a strip of paving flags running diagonally through the cobbles tells one which way to go. Similar examples abound of routes for wheeled traffic through pedestrian precincts. They then also function as in Hazard below. There are two things to note about their design. First, to decide on their route one must simply join the two points which are to be connected and draw a beeline between them; otherwise as direction finders they will not be sufficiently telling and urgent. Secondly, the change of colour and texture of material must be subtle, not violent. One is aiming to preserve the unity of the surface as a whole, so the contrasts must be quiet if they are not to be disrupting. The shapes left may be odd and so should not be accentuated. This also means that the background material will probably have to be rather small scale or it will be difficult to work into the acute angles and odd corners that this treatment necessarily creates. It will also be seen how fatal a last minute substitution of grass for the hard background material would be. Both visually and practically, paths across grass will need additional and different consideration.

1.1.3 Repose

To provide a sense of repose; an area of neutral, non-directional paving has the effect of halting people. In a garden that might be at a sheltered place to sit, in a town square it might be at a place to stop and gossip or buy a paper, or it may be somewhere to have a drink and gaze at a view, or rest in the shade of a tree.

1.1.4 Hazard

To provide an indication of hazard by change of material; this method, if it is used consistently, becomes a language on its own. A traditional example is the use of setts in a flag pavement where a private road or drive crosses it to reach the street. Though their origin may have been simply to provide a tougher material for carriage wheels, their presence makes pedestrians immediately aware that they are stepping off their own sanctuary – the footpath – onto a surface where vehicles may be expected. Similar use is made of cobbles between paving slabs and grass. Here they discourage people from wandering off the path, or cutting corners. A band of setts at the junction of a private and public road will not only demark ownership, but will also slow the traffic at the junction.

1.1.5   Scale

To reduce scale; this is the most difficult function to handle without making some blunder. Because there are only psychological reasons for changing materials, and the designer is not guided by practicalities, changes can easily look forced. On analysing some examples it is found that a material which is used in traditional work for a particular purpose, has been introduced simply to give pattern. Brick or slab strips in paving are an example. Originally used as a drainage channel where two sloping areas of paving meet, they are now used as dividing lines only. The subtle lines, which were entirely acceptable when a response to practical necessity, can appear either pretentious or timid (and sometimes, somewhat paradoxically, both simultaneously) when used solely as a visual device. Timidity can also sometimes be seen in another form – an assumption that any area of paving, even if only of moderate extent, has to be broken-up visually. Whilst this might be excused as a natural reaction to the seas of tarmacadam and concrete that are too often seen, if one looks at obviously successful examples it is often remarkable how simple they are. For example the whole of the middle of Sloane Square in London – 2033 m^2 (2430 sq. yd) – is virtually paved as one area with very generous falls to the road so that there is no need for dividing up by channels; trees break down the scale, both by the vertical pattern of their trunks and by the pattern thrown by the shadows of their branches. It is in this respect, if not architecturally speaking, that it is one of the most pleasant squares in London.

A pattern of paving slabs which have scale in themselves, needs little sub-division. Macadam or *in situ* concrete surfaces may need to be visually subdivided. But, perhaps, consider first whether a wearing course of uncoated chippings might not give the macadam gentle colour and fine textural interest such that the very area has positive impact rather than being a daunting extent of matt-blackness. Consider also whether the expansion joints which are a technical necessity of *in situ* concrete paving, might not be organized in such a way as to have visual benefits as well.

When a change of material is introduced simply for the sake of scale, the change should be rather quiet if it is not to make the paving absurdly important. Paving is, when all is said and done, almost always no more than a background and should not aggressively proclaim its presence. Slight changes of colour and/or texture within the same type of material are often more successful than a change in actual material. But practical needs can be exploited for aesthetic effect. Surface drainage – a matter where one suspects designers are often plain lazy – is a case in point. It can provide an admirable practical reason for breaking down scale by the use of flat or dished channels, simultaneously giving the paving a trim appearance and a manageable scale; furthermore there is also the great advantage that they actually drain after rain!

1.1.6 Use

Ownership and use are closely related to scale. That parade grounds demand a different scale from garden paths is self-evident; the public military character of the first is so opposed to the informal domestic nature of the second. But there are also less easily defined ways in which scale can indicate such things as privacy or rights of ownership. A change from the scale of paving flags on a public pavement can suggest private property without any change of level or fencing. This device is often used at entrances to hotels and shops, or where a café or restaurant puts tables on the pavement. It has the effect of making the public realize they are on private property, which they are welcome to use as clients of the establishment, but that they are otherwise expected to move on. Generally, but not always, smaller scale paving seems to indicate private ownership; larger scale, public property.

1.1.7 Character

To reinforce the character of a particular place; paved surfaces, and particularly their edge treatments, can have enormous effect on the character of the place of which they are a part (see also section 1.17). Before deciding how to pave or detail a particular place, the desired character must be identified or, if that place is wholly new, decided upon. Is it to be formal or informal, urban or rural, crisp and precise or mossy and blurred, rugged or sophisticated? Whatever its character (and no words can ever exactly carry the synthesis in these cases) the surface must be appropriately durable and must improve with weathering. Immense variety is obtainable within quite a limited range of basic materials. From the precision of blue pavers at one end of the scale through to lichen covered stock bricks at the other; or from the rigid outline of a macadam surface edged with upstand kerbs to the informality of the same material when kerbless or a lane with grass growing down the middle. However, one must always have a clear sense of the objective before starting.

All materials must fulfil the first provision on this list – that is they must provide a useful pavement; but in respect of the other less tangible functions, no material can be said to have a single unalterable character. Its effect depends upon its juxtaposition with other materials. Granite setts, for example, provide a rougher surface texture and a smaller scale than most other materials, so they are often used as hazards, but they might elsewhere be used as direction finders within cobbles, which are even more awkward to walk on. Again, cobbles themselves are sometimes used for footpath material – they are at least easier to walk on than soft and/or muddy earth.

It must be emphasized again that really successful paving reinforces the latent or existing character already dictated by the space to be paved. In this way it can make or mar a scheme. It is not suggested that the paving alone will create character, although in certain cases this too is possible. It can be used to bring unity into what might otherwise be too diverse a design, or to give character to some rather nebulous area which needs a common background or idea. If the same material is used for all pavings, steps, plinths to buildings and free-standing walls, a motley group of buildings can be given unity and the material, be it brick, stone, concrete or whatever, appears rather as a layer in rocky country, the various levels corresponding to different strata. A sculptural effect is achieved, as though the paved area had been carved out of the whole.

1.2   PRINCIPAL HIGHWAY ENGINEERING TERMS

Each of these items and its function are described in the following text. This glossary is included here for quick reference only.

1.2.1   Pavement

(a) A general term for a paved surface (British readers should note the potential confusion in that the word is often used colloquially to refer to a particular kind of surface which, elsewhere in the world might be called a sidewalk).
(b) A term applied specifically to the whole construction of a road.

1.2.2   Flexible structures

A form of road construction which for the purposes of design is assumed to have no tensile strength. Such roads are built up of layers of granular materials. In the upper layers there is frequently a binding agent, most often bitumen or tar.

1.2.3   Rigid structures

A form of road construction in which, for the purposes of design, the tensile strength is taken into account. Reinforced *in situ* concrete is the best (indeed for most practical purposes, the only) example. (But that is not to say that all pavings using *in situ* concrete are rigid structures – soil-cement and lean-concrete bases, for example, are flexible.)

1.2.4 Subgrade

The soil (and sometimes, fill) which carries the road.

1.2.5 Formation

The surface of the subgrade in its final form after completion of all preparatory earth works.

1.2.6 Sub-base

A layer of material (which is not always needed) between the formation and the road base.

1.2.7 Base

The main load-spreading component of the road, which provides the principal support to the surfacing.

1.2.8 Surfacing

Waterproofs the road and provides a non-skid running surface. Sometimes provided in two layers, the base course and wearing course.

1.2.9 Surface dressing

Usually consisting of small stone aggregates spread on or in a thin layer of binder, but not always included.

1.2.10 Macadam

The stones which, in the original roads designed by Macadam held together because the smaller interlocked within the interstices between the larger, and which now (particularly in the base and surface layers) are often bound together with a binder.

1.2.11 Binder

A viscous material which binds the aggregate together. It may be tar, bitumen or a proprietary mix. Pigmented binders are available which can radically change surface appearance.

1.3 GENERIC TYPES OF PAVING

The above glossary might give the impression, (at least through implication) that there are only two kinds of road base construction – reinforced *in situ* concrete which embraces all rigid structures, and tar or bitumen bound macadam which embraces all flexible forms. It is true that all rigid structures involve the use of reinforced concrete, but even then the range of surface finishes that can be considered, and the possibility of using a reinforced concrete slab as the base to a number of surfaces which in themselves might not have necessary strength, gives this approach considerable versatility. However, tar or bitumen bound macadam is not the only generic form of flexible structures. As with Macadam's original design, graded aggregates can be used without any binder – graded stone chippings or naturally occurring gravels are probably the commonest construction for private drives. Further, as has been said, graded and consolidated aggregates that are bound with cement but without steel reinforcement, are also flexible structures.

There is yet another form of flexible construction which is becoming increasingly common – indeed could well become the most common design approach to lightly trafficked roads. Both clay and concrete bricks can be vibrated into a sand bed overlying the road base. There is no viscous binder in this form of construction for the mechanical bond between the units (amplified by the sand which works its way up into the joints between them) is sufficient to hold them in place. For more highly stressed roads irregular forms of the bricks can be used, giving yet higher degrees of mechanical bond.

1.4 PAVEMENTS FOR WHEELED TRAFFIC ONLY

These notes are concerned with the design and construction of minor private roads and other areas carrying wheeled traffic, but observations included might well apply to certain aspects of highway design.

It must be said at the outset that in many circumstances the landscape designer may wish to call on the expert advice of a civil engineer with work of this kind – these notes make no attempt to offer anything more than very basic information. But the detailed design of any road, however minor, has such a drastic effect upon the landscape of which it is a part, that it deserves the special attention of the person responsible for the landscape as a whole. These notes will help the landscape designer to work effectively with relevant specialists.

1.4.1 Details and character

Roads must be counted among the most dramatic and romantic things constructed by man – the Royal Road of the Persians, the gigantic network of the Romans, and even the ancient British tracks, have the most evocative names.

This book, concerned as it is with the detail design, is no place to discuss the principles of planning and routeing a new road. Ironically enough this major problem is now often adequately, and sometimes splendidly, solved only to be wrecked by fussy detail. There seems to be a greater tendency to become aesthetically unstuck on matters of detail than on broader issues; it is detail which consolidates and often creates character. Two roads of identical dimensions and otherwise similar design can, for instance, be given utterly different characters by their edge treatments. The most obvious example of this is the pitiful result which occurs when kerbs have been introduced in rural areas; this is done every day in the name of improvement. The road is no longer a part of the country through which it travels; it is divorced from it. The hard line and sudden change of plane at the kerb make the road a foreigner, an intruder from the suburbs, no longer integrated with its surroundings. The kerb carries the tang of urban sprawl into the heart of the country as surely as (and far more permanently than) litter or the blare of a car radio on a summer afternoon. All these are intruders and by their unnecessary presence spread the uniformity of subtopia yet further.

Similarly, the surface colour or texture of a road may be entirely foreign to its surroundings. As communications improve it becomes increasingly easy to move aggregates around the country. No longer is it probable that uncoated surface dressings (say) have been quarried from the same county as that of which the road is a part. For example, strident red granite may streak across a gritstone moor – if those with a broader view have no influence.

One cannot discuss the pros and cons of such details or materials without knowledge of their precise function: therefore the notes which follow are intended both as a key to the details of design which confirm the character of the road, and as an elementary introduction to construction.

1.4.2 Responsibility for construction

It is assumed that most public and larger private roads will be designed by a road engineer; but within building contracts there are often cases of small private roads which hardly warrant such attention. They then form a part of the external works contract and become the responsibility of an architect, landscape architect, land agent or owner. A situation then arises, typical of others in practice, when the non-specialist must be able to sift the advice and respond to the arguments of sub-contractors, in an area where there may be little experience or training. This section is not intended to bypass the services of the specialist, be that engineer or contractor, but to assist the non-specialist in making greater use of the services available, first by putting the designer in a position to discuss and make decisions on the various problems of design and construction, and secondly by briefly referring to organizations and publications that offer further help.

1.4.3 Local advice

The Transport and Road Research Laboratory advises that the first thing to do when confronted with a problem of road design is to discover the local current practice. The reason is that in road construction the use of locally available materials is still extremely important economically. Whilst vast engineering projects might consider transporting slate waste from North Wales to the Thames estuary it is only because the quantities required override other economic constraints. The materials are bulky, a lot are needed and transport is expensive. Therefore the type of construction decided upon for small works may well be decided by the materials that are reasonably to hand. The District Authority's engineer, surveyor or building inspector should be able to give sound advice both on local natural materials and on those resulting from local industry. For example, it might be more economic to build an *in situ* concrete road (which when all else is equal – which it generally isn't – would be more expensive than a flexible pavement for light roads) because the site is immediately adjacent to particularly suitable sand or gravel pits or a ready-mix works. Conversely, there might be arguments for a macadam road if the local quarry is producing rock that is better for a base course than for concrete aggregate, or if there are local demolitions happening at that time producing broken concrete or brick which can be ideal below light roads. Local advice should also help identify reputable contractors able to carry out work of that kind and extent.

The form of contract which it proves appropriate to sign will vary with both the value of the work and whether or not it is the whole of the work. But whatever the form, clauses can be included which make the contractor responsible for technical aspects of the design and for maintaining the road for one or two years, provided that stated traffic intensities are not exceeded.

1.4.4 Further sources of information

Roads Notes and other publications of the Transport and Road Research Laboratory (TRRL) are referred to in detail later in this section. They contain much useful information, but are primarily intended for highway engineers. This must be remembered when reference is made, for the loads on private roads are generally very small compared with those on the average highway. Whilst this may be noted, one is apt to forget it later and be carried away either by enthusiasm for a new subject or because of a natural tendency to over-design in cases of uncertainty. It is not unknown for a service drive to have a cross-section which would not disgrace a motorway. The TRRL is not in a position to welcome specific questions of this kind since it is a government organization engaged on scientific research. However, there are other organizations who are both ready and able to advise. As they deal with all kinds of pavement they are listed in section 1.5.

1.5 TRADE ASSOCIATIONS

When it has been decided which type of construction is likely to prove most suited to a particular job, there are several trade associations who will advise, each in its specialized field.

Asphalt and Coated Macadam Association. Will advise on tar and bituminous materials and deals with all parts of flexible road construction but is particularly helpful with questions of surfacing, both hot and cold rolled.
(156 Buckingham Palace Road, London, SW1W 9TR, 01–730–8194)

Ready Mix Concrete Association. An association of the many firms about the country which make ready mixed concrete under conditions of close control which ensure high quality. The association will give the name of the member firm closest to any particular site, and that firm will advise on the specification of the required concrete when the job which it has to do is described. Ready mix concrete is of particular value on restricted sites where lack of space makes the handling and mixing of materials difficult, and with projects where the quantity of concrete needed is greater than could be managed by hand or with a small machine but less than would justify setting up a mixing plant on site.
(19 The Crescent, Ilford, Essex, 01–592–0297)

British Cement Association. This association runs a large research and educational establishment and has done a great deal of useful work on the appearance as well as the construction of roads and pavements. It does not try to sell cement to the exclusion of other materials.
(Wexham Springs, Slough, SL3 6PL, 02816–2727)

The Concrete Block Paving Association (Interpave) and the *National Paving and Kerb Association.* Two different associations which share the same address and telephone number with others that also have an interest in precast concrete. Whilst only the most general technical questions could be answered by the associations themselves, inquirers will be directed to those member firms best able to help in a particular situation. Lists of all members are available together with literature describing the generic products.
(60 Charles Street, Leicester, LE1 1FB, 0533–536161)

British Aggregate Construction Materials Industries is the national trade federation for the aggregate, coated materials and ready mixed concrete industries. It includes in its members companies producing roadstone, sand, gravel, asphalt and coated macadam, as well as flexible paving contractors. It has a number of groups (of which ACMA, mentioned first above, is one) including the Plant and Equipment Group (companies manufacturing construction and quarrying plant and equipment) and the Refined Bitumen Association (producers of petroleum bitumen).
(156 Buckingham Palace Road, London, SW1W 9TR, 01–730–8194)

Transport and Road Research Laboratory. Not a trade association but a government research station, included here so as conveniently to give the address.
(Old Wokingham Road, Crowthorne, Berks. RG1 6AU, 03446–3131)

British Standards Institution. Again, not a trade association but included here because of its importance in producing standards affecting road making materials and their use; and as a convenient place to include the address.
(British Standards Institution, 2 Park Street, London, W1A 2BS; also Linford Wood, Milton Keynes, MK14 6LE)

1.6   GENERAL DESIGN QUESTIONS

As in all design the first thing to do is to discover what the questions are and to decide what particular broad answer best integrates the aesthetic and practical requirements of the case – which is often much easier said than done. The questions might be listed as follows.

1.6.1   Planning

That is the road both as a practical route and in relation to the landscape of which it will form a part. Overall line, widths, location and form of junctions, relationships with associated land-form, planting, footpaths, walls, steps and buildings. The theory which informs many of these questions is way beyond the scope of this book, but some detail aspects will be discussed.

1.6.2   Types of surface

Some aesthetic considerations have already been explored, colour and texture in particular. To these must be added matters of character of line (some forms of construction dictate a regular overall geometry whilst others are better able to accommodate to subtle curves, width variations, or whatever), the nature of the edge details demanded by each form of construction, demands for, or capacities to absorb, gully gratings, inspection chamber covers, etc. To these aesthetic considerations must be added questions about the practicalities of the surface itself. What wear and tear must be resisted, will the traffic simply be driving through (giving more wear than tear) or will there be the scuffing effects of slow moving but manoeuvring vehicles (more tear than wear), do oil drips need to be anticipated (an aesthetic problem needing practical solution), are there other matters (such as draining surface water) which are constrained in such a way as to, in turn, limit choices about surfacings?

1.6.3   Structural requirements

What loading will there be on the pavement, what is the capacity of the substrate to resist that loading, does the substrate have mechanical properties which might make some forms of construction more difficult to achieve, what is the groundwater level, does it fluctuate, are circumstances such as to preclude certain strategies?

1.6.4   Budget

What are the local comparative costs of different materials, what sums are available for the work, can a low initial budget be compensated by funds available for maintenance and repair, or must (relatively) high initial cost be accepted so as to give a road with a long, maintenance-free life that avoids future disruption?

To summarize; it is most likely to be the nature of the subsoil and the load the road has to carry, together with the surface required and the availability of materials at an acceptable price, that decide choice of structure – whilst it is the character of the landscape (or townscape) which has to be created or preserved which determines tactical details within that broader decision.

Before describing the various types of pavement in detail, problems concerning all types will be discussed – that is those concerning the soil under the pavement, the loading, and (briefly) ways in which to find out which materials are likely to prove most economic for a particular project.

1.6.5   The subgrade

(i) *Strength.* It is the soil under the pavement that carries the load; the pavement itself distributes the load over the subgrade. Figure 1.1 illustrates the distinction between pavement, subgrade and formation. Macadam said 'Roads can never be rendered perfectly secure until the following principles are fully understood, admitted, and acted upon, namely that it is the native soil which really supports the weight of the traffic, and that while the soil is in a dry state it will carry any weight without sinking'.

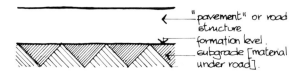

1.1 *The pavement distributes the load imposed upon the road over the subgrade. The prepared surface of the subgrade is the formation.*

The first thing then to be decided is the strength of the soil which is to carry the road. The road engineer, working on a far larger scale than the jobs envisaged here, will have soil samples taken and tested in a laboratory. Even on a small job the cost of taking the samples (by boring) and having them tested would be well in excess of £1000, and one cannot expect a client to spend such a sum for a minor private road or drive. Much better to expend just a part of that sum putting down additional base material which would give a factor of safety against both the designer's lack of precise information and some unforeseen load which the drive was never expected to carry. Laboratory tests are essential for the public road where their cost is a small percentage of the whole and where precise information is essential to ensure high-class work.

Even if the client could afford the tests, the answers provided, unless they corresponded closely with conclusions reached on visual inspection, might cause an awkward division of responsibility. The subcontractor, with considerable experience of road building and of the climatic and human failings involved, might well wish to use more material than the tests suggested. Under such circumstances he is unlikely to want to take responsibility for the road unless his advice is followed. Laboratory tests can, after all, only show what load the soil should be able to carry; they give no guarantee that the road will satisfactorily distribute that load on to the soil.

When laboratory tests are not to be used, trial holes should be dug to a depth of at least 900 mm (3'0") below formation level (having taken an educated guess as to how far below existing ground this is likely to be) and a note taken not only of the soil type(s) found but also of the rapidity with which water drains through it and the level of the water table at its worst. Soil strengths vary considerably with water content and so this last is of critical importance.

Trial holes dug for roads have similar practical snags to those which are sunk for buildings, with the addition that a road, since it stretches over a greater length of ground, may have a greater variety of soil types under it. Decisions about the number of holes dug must be guided by what is discovered from the first of them, and by local knowledge. In some cases it may be known that very stable conditions will be found fairly quickly, and so the trials are needed simply to establish precise depths. In other cases all that might be anticipated is that great variety will be found, and then many more holes will be required – perhaps in order to establish whether a road of any kind could be built.

Trial holes should not be filled in if that can possibly be avoided for it is almost certain that further inspections will be desirable, but they should always be fenced if there is any danger of stock (or people) falling in.

(ii) Soil types. If the soil types vary much, let common sense prevail. It will almost certainly be cheaper to give the whole of the road the foundation needed in its worst part than to be constantly changing its design ideally to suit conditions found in each small area. It is difficult without some experience, to recognize those soils which the experts consider to be **normal** or **very stable** or, much more import-

antly, those which are **very susceptible to non-uniform movement**. All these expressions crop up in the technical literature. The following list is given in order of strength, always assuming that compaction and water content are at their optimum for the material concerned.

Very susceptible to non-uniform movement
    galt clay
    heavy clay
    subgrades with pockets of peat within 4.5 m (15'0") of the surface
Normal
    lighter clays, lias clays
    alluvial deposits, silty clays
    sandy clays
    fine sands
    sandy gravel
    gravel
Very stable
    solid rock
    well-graded compacted gravel with a CBR* of at least 100% (*CBR = Californian Bearing Ratio; section 1.7.1).

(iii) Water in subgrade. Excess water in the subgrade reduces its bearing capacity and may cause frost-heave (deformation of the surface because ground water has expanded when turning to ice) in really cold weather, particularly in chalk soils. Frost may also disintegrate the road itself if its nature allows it to hold water which then freezes. Therefore one must beware of rain getting in from the top if it cannot escape freely below, water seeping in from the sides, and seasonal changes in the level of the water table. In the first case the road must be protected by a waterproof surfacing and in the second and third land drains and/or ditches must lower the water table to an acceptable level.

Where it is critical to prevent rain percolating through a pavement then a form of construction must be used which has a waterproof surface. However, waterproof is always a relative term and so it is also important to give the road a cross-sectional profile which is either cambered (with the middle higher than the edges) or has a cross fall (one edge higher than the other) of between 1:80 and 1:120. Such falls have the distinct additional advantage that puddles will not form. Indeed falls are often given to roads that do not absolutely need protection from percolation simply to avoid puddles.

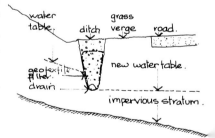

(iv) Drainage of subsoil. The object of drainage of the soil under the road is to keep the moisture content reasonably constant, for if that stays constant then so will its load-bearing capacity. A water table which rises in winter, when it might freeze, to within 900 mm (3'0") of the formation level, should be lowered. In country areas this can be done with open-sided ditches or French drains (Figure 1.2). The ditches are dug down to the level to which the water table is to be

1.2 *Excess water in the subgrade reduces its load-bearing capacity. Ground water levels can be lowered by constructing a land drain in the verge.*

reduced. These are then filled with free-draining material (say gravel, clinker or stone chippings between 20 and 75 mm (¾" – 3") to 150 mm (6") of finished ground level. A filter blanket (one or other of the proprietary geotextiles) is then used to minimize the rate at which fine material might wash in and inhibit drainage, and ground brought up to finished levels with topsoil. Where large ground-water flows are feared or heavy rain anticipated (particularly when the French drain is also disposing of surface water run-off from the road itself) then a land drain of porous clay or concrete with butt joints, or slotted plastic pipe of 75–100 mm diameter (3"–4") should be laid at the bottom of the trench before it is back-filled, and taken away to discharge into a soak-away (a hole with its base above ground water level, with a volume four times greater than the volume of water it is anticipated will drain into it, which is back-filled and covered as described for the French drain itself). Soak-aways must be located where they will not present difficulties and where one can be certain that the water will drain from the road to the soak-away and not *vice versa*. Clearly the water can be taken into a mains surface water drainage system if one is available.

Arrangements of field drains below the road are not recommended for it needs but one abnormal load to produce a situation which is worse than if those drains had never been installed. In towns the matter is best discussed with the local authority, for the opportunities to locate soak-aways may well be less but the possibilities of connecting with a surface water drainage system may well be more.

(v) Preparation of the formation. The formation is the name for the surface of the material on which the road is constructed. All organic topsoil must be removed and tree roots grubbed up, for the nature of decomposition will change once air is largely excluded and there would be settlement where had once been organic material. The depth of topsoil to be removed will vary from site to site, but a colour change will always indicate its depth. (If in doubt, take out more!) Further excavation may be necessary to reach formation level if depth of topsoil is less than the necessary depth of construction. Conversely, the subsoil may have to be made up (particularly if finished road level is to be higher than original ground level at any point) or made good where there are local deficiencies in that subsoil (naturally occurring soft spots, unconsolidated fill in previously excavated trenches or dumps, or bad practice during excavations having produced wheel ruts or depressions which have subsequently puddled after rain). Any saturated areas should be excavated to a good hard subsoil. Levels can be made up or made good with material which has been excavated elsewhere on site, but with the strict proviso that such material, after consolidation, must perform structurally in the same way as the neighbouring undisturbed ground. Alternatively, levels can be made up with any suitable base material such as described below. The loose fill should be spread in layers not exceeding 225 mm

(9″) and then consolidated with a heavy 8–12 tonne (8–12 ton) roller, or its equivalent. If such a weight of roller is not available, or is itself physically larger than the site can easily accommodate, then backfill should be in 150 mm (6″) or even 100 mm (4″) layers to ensure that a light 0.5 – 0.75 tonne (10–15 cwt) roller, or its vibrating plate equivalent, will achieve necessary compaction. Clearly, the fill material should not be saturated with water for then it would perform no better than that which has been removed. If it is not entirely dry then it would be better practice to consolidate with a light roller in thin layers.

The formation as a whole should be well rolled and consolidated to the identical profile as that required of the finished surface. The subsoil and/or fill below the formation does not have to be perfectly dry but should be drier than is likely to be the case after the road is built and in use. Rolling will not reduce the moisture content of a soil but if it has been compacted to a high density when relatively dry it will take up less water subsequently. To achieve a high bearing capacity in a soil both a low moisture content and a high density are required. But heavy clays should not be rolled; they cannot be compacted beyond their normal state, and rolling only tends to remould and thus weaken them.

In a climate where rain is a probability some time between excavation and construction it will generally pay to lay a temporary waterproof surface over the formation; for example, a sprayed bituminous binder at a rate of 5 litres/m^2 (1 gal./sq. yd) which is then dusted with sand so that it can be walked (but not driven) over prior to construction starting. Care must be taken to ensure that any water that might collect can drain away, otherwise the sealed formation may turn the excavation into a swimming pool. (Good practice minimizes the time interval between excavation and construction.)

(vi) Excavated material. In order to avoid heavy (and from the contractor's point of view, justifiable) extra costs, it is worth planning beforehand where the separate dumps of excavated top- and sub-soils are to be, or whether they are to be spread or carted away as work proceeds and, if so, where. Such planning will avoid double handling which is always a waste of both time and money, or, if it cannot be avoided then, at least, it can be accounted for before work starts. If material is to be taken off site then its destination should be named, either by the designer at tendering stage or, if that has not proved possible, then by the contractor before any work starts. Otherwise there is always a possibility that the material gets dumped around the first convenient corner; a practice which is, at best, embarrassing for both designer and client.

1.6.6 Loading

The depth of construction necessary below the surfacing of a road is conditioned by the loads that will be imposed by the vehicles using that road and by the load-bearing capacity of the subgrade. That is, the imposed load (which in itself is almost certainly greater than bearing capacity) must be spread as the forces pass down through the structure of the road so that the load per unit area at formation level is no greater than the subgrade's bearing capacity. So, the greater the difference between imposed loads and subgrade bearing capacity, the greater will be the necessary depth of construction for any particular structure.

It should be noted that the matter is further affected by the number of vehicles using the road. All roads deform fractionally whenever they are loaded. If the road is properly designed that deformation will be purely temporary and it will recover again as soon as the load has passed any particular point. But, these deformations and recoveries will cause frictions within and between the aggregates in the road and, in time, it will weaken and wear out. The busier a road the greater must be its depth of construction so as to prolong the period before that weakening reaches unacceptable levels. However, the most lightly trafficked road considered in the various 'Road Notes' is 0–15 commercial vehicles (excluding light vans) per hour. It is not, then, very likely that construction depths of private drives would have to be increased for this reason but this point might well be raised in discussion with a road engineer should the designer be involved, say, with the external works around a hypermarket.

It should also be noted that it is the wheel load rather than the pay load of a vehicle which counts. For example, about two-thirds of the load of a lorry is taken by its rear wheels, and so a six tonne (six ton) lorry with four rear wheels will transmit similar loads to a road surface as will a three tonne (three ton) lorry with two rear wheels.

Whilst tar-bound macadam, reinforced concrete and interlocking pavements will be considered later in some detail, there is value in summarizing here the principal mechanical differences between them and the effects of those differences on depths of construction. As has been said, flexible structures have no tensile strength. (To be more precise, they have a little but it is too little to be useful and would, in any event, be very difficult to quantify.) So, a flexible road can only spread forces through its structure from a point load on its surface, in the broad form of a cone. That is, the load on one piece of aggregate will be transmitted to the small number of others that are immediately below it, and these in turn will pass those forces on to the small number below them, with the number of pieces of aggregate sharing the load steadily increasing with increased depth. The load imposed upon any one piece of aggregate will decrease with increased depth, for a larger number will be sharing the total load. So, given that the overall depth of construction is correctly judged, by the time formation is reached the load on each single piece of aggregate is no greater than can be resisted by that small part of the subgrade immediately below.

Again as has been said, rigid structures do have tensile strength, and in both useful and predictable quantity. The steel reinforcing mesh placed within the lower part of an *in situ* concrete road will allow the forces from a point load to be spread laterally very rapidly within the depth of that slab. The reinforcement gives the slab a resistance to bending (the deformation described above). If it could bend then the loads would be transmitted locally to that part of the formation immediately below the bend. As it cannot bend (or rather, can only bend a very little), those loads are spread over a much wider area. Both the amount of bend that would be induced by particular loads, and the depth of slab and amount of reinforcement necessary to resist those bends, can be calculated accurately by a structural engineer. In all cases other than those where the subgrade has a very high bearing capacity and he anticipated loads are very light, the depth of construction needed for a rigid structure will be less than that needed for a flexible road. Figure 1.3 indicates how loads are transferred through flexible and rigid road structures.

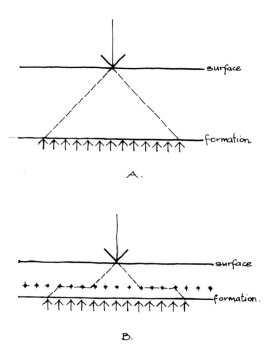

1.3 *Loads are transferred through a flexible structure in the broad form of a cone (A) whilst the reinforcement in a rigid structure (B) gives tensile strength which allows loads to be spread over a similar area, but with less depth of construction.*

1.6.7 Comparative costs

Since the cost of all hard surfaces, but of material for roads in particular, is much a matter of transport and local availability, including even the use of 'waste' materials, the use of local materials is all-important. Local advice must be sought; it would be unrealistic to pretend to be able to go much further here. The detailed conditions in your locality would almost certainly prove to be an exception, and ten minutes spent with the local District Engineer or Surveyor would be of more value than anything that might be written here. He or she should be contacted early in the project, since the advice given will restrict the otherwise overwhelming list of materials to a few economic possibilities.

What can be said is, everything else being equal (and it never is), then the excavation costs for a bound macadam road are likely to be higher than for reinforced concrete construction (for they are likely to be deeper). But the labour and material costs for the former are likely to be lower to a more or less compensating degree. Only very rarely would the overall cost of an interlocking block pavement be lower than the other principal forms, but the number of such pavements now being laid shows that the question of appropriate structure is not, and cannot be, judged on costs alone.

More precise information on costs can be found in the latest edition of the *Landscape Price Book* published and regularly up-dated by E. & F.N. Spon. When using that book make sure that all costs for labours and materials are taken into account (excavation costs, the costs of various depths of base and surfacing costs, for example are given in different sub-sections) and don't forget the costs for drainage and kerbs, for the relative costs of these will also have a relevance.

1.7 TAR, BITUMEN OR HOT ASPHALT BOUND FLEXIBLE PAVEMENTS

A flexible pavement is a structure made up of layers of thoroughly compacted material which will spread the traffic load on the subgrade below. The thickness of that construction will depend upon the strength of the subgrade, the traffic load and the strength of the materials used. Each of these layers (illustrated in Figure 1.4) has its own special function.

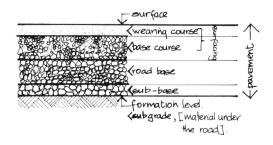

1.4 *A bound macadam flexible road structure is built up in layers; the depths of each of these will vary with circumstances.*

(i) *Sub-base.* A layer of fill used when the theoretical minimum depth of construction is less than the depth that will actually be needed in order to adjust levels or whatever. The material used will be very similar to that for the base layer, but it need not be as strong for forces within it will be less.

(ii)   Road base. The main load-spreading component of the road construction. This itself might be laid in more than one layer if calculations show that it needs to be more than 100–150 mm (4–6") thick when only a light roller is available or more than 225 mm (9") thick when a heavy roller will be used (section 1.6.5).

(iii)   Surfacing. Waterproofs the road and provides a non-skid running surface. This may be only one layer (particularly on lightly trafficked roads) but may have both a base course and a wearing course. The theory being that the wearing course is an additional layer and will be repaired before the amount of wear starts to erode the base course, and so the road proper will never have less than its designed depth and strength.

1.7.1   Method of design

There are several ways of designing a flexible pavement none of which, from the standpoint of pavements discussed here, is entirely satisfactory in itself.

(i)   Scientific methods. One of the scientific methods developed for large road schemes can be used. The best known of these is the Californian Bearing Ratio (CBR) method. This entails full use of scientifically made soil tests. This would be an uneconomic method for the scale of project envisaged here for the costs of the tests might well be in excess of £1000. However, the nature of the tests and the conclusions that can be drawn from them are briefly described in section 1.7.2, for a broad understanding would be useful when working with a specialist road designer.

(ii)   Designers' experience. Plus, possibly, a good factor of safety. The advantage of this approach is that clients do at least get material results instead of theory for their money. However, this can hardly be considered as satisfactory advice where experience is thin.

(iii)   Subcontractors' experience. If a reputable firm is employed on a design and build basis, there is no reason why this should not be satisfactory. But everything will depend upon the firm. This method is only really acceptable if (a) you are spending your own money and not that of a client, or (b) you have sound reasons to have confidence in the contractor. But whatever the circumstances, some understanding will be necessary both when selecting a contractor (be that by competitive tender or negotiation) and when supervising the work.

An amalgamation of all three of these approaches must be a reasonable solution, and should help the inexperienced to judge and select. Trial holes will have to be dug and the subsoil type(s) identified. The loads which will be induced by the traffic will have to be decided. With this information the local authority's engineer or surveyor should be able to supply information on local practice and a list of local contractors who might be interested in work of that kind and size. The information about local practice should have reduced what might have originally appeared to be a bewilderingly long list of possible materials, to one of more manageable length in which there can be confidence that there is basic economy.

It must then be decided whether or not the contract is to be let as a whole or in two parts. Not infrequently when there is already a general contractor on site, he is asked to undertake excavations, prepare formation, and lay and consolidate the sub-base and base, on to which a specialist firm will put the surfacing. In such circumstances it is most important that the specialist firm is party to the design decisions for the whole of the construction and has opportunities to inspect the general contractor's work as it progresses.

1.7.2 Californian Bearing Ratio

This is an empirical method based on the experience of the thickness of material required to enable a road to carry various loads of traffic on different types of subgrade. By this method, the bearing capacity of the soil, the sub-base, base and surfacing materials (and their thicknesses) can be determined. The strength of the subsoil is measured by penetration tests when the moisture content of that subsoil is at the highest likely during the life of the road (that is, when the soil is at its weakest). The soil is given a CBR figure, which is expressed as a percentage of the necessary strength of the surfacing (i.e. 100%). Once this and traffic load is known, the necessary depth of construction can be read from the graph given in Figure 1.5.

Further, the CBRs for the various consolidated materials of which the road is/might be constructed, are also known, and so the depths of the various layers (each with a decreasing CBR) which collectively make up overall depth, can also be found from the graph. For example, assuming that the trial holes have disclosed a light clay subsoil (CBR 4%) and the road is to be designed for very few commercial vehicles (curve A), then the overall depth of construction will need to be 257 mm (10"). A desire to give a factor of safety might well round up that figure to 300 mm (1°0").

Assuming also that investigations of economically available materials suggest that base material with a CBR of 80% when consolidated, and sub-base material with a CBR of 15% when consolidated, would best be used, the graph shows that 125 mm (5") of cover is needed over the sub-base and 50 mm (2") over the base.

Decisions about the colour and texture of the surfacing (CBR 100%) must then be taken and advice as to the depth of such a surfacing sought, say 50 mm (2").

So overall depth of construction equals 300 mm (12") of which the sub-base must have a thickness of 175 mm (300 mm minus 125 mm) (12" minus 5"), the base a thickness of 75 mm (125 mm minus 50 mm)

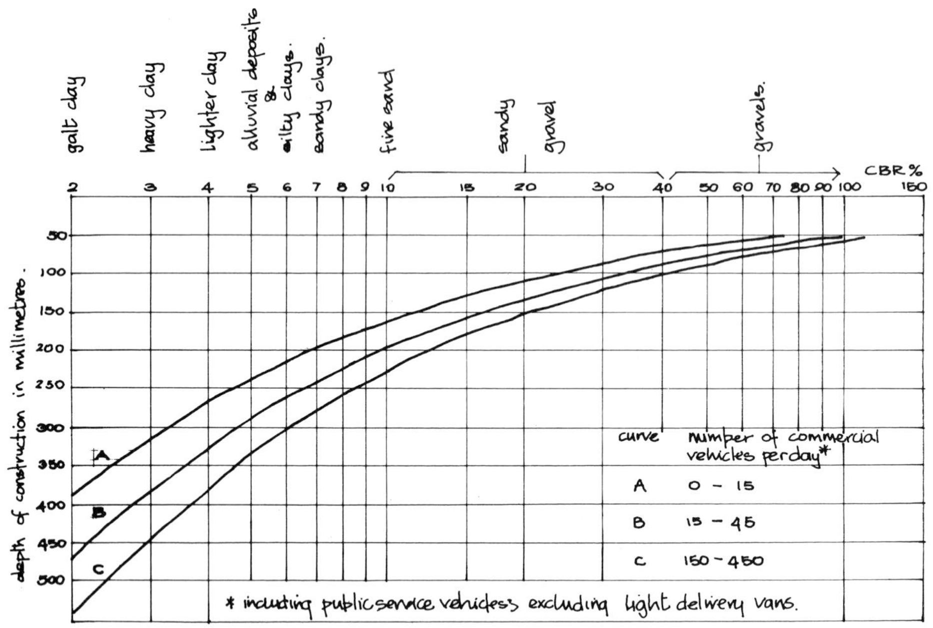
galt clay
heavy clay
lighter clay
alluvial deposits &
silty clays.
sandy clays.
fine sand
sandy
gravel
gravels.
CBR %
2 3 4 5 6 7 8 9 10 15 20 30 40 50 60 70 80 90 100 150
depth of construction in millimetres.
50 100 150 200 250 300 350 400 450 500
A
B
C
curve
number of commercial vehicles per day*
A 0 – 15
B 15 – 45
C 150 – 450
* including public service vehicles excluding light delivery vans.

1.5 *The Californian Bearing Ratio (CBR) curves for varying intensities of traffic, and for different types of subgrade, aid the design of bound macadam roads.*

(5″ minus 2″) and the surfacing a thickness of 50 mm (2″). Again a decision to be safe might well be taken and the road designed and constructed with a 150 mm (6″) sub-base, a 100 mm (4″) base and a 50 mm (2″) surfacing. Figure 1.6 takes the CBR graph and superimposes upon it the results of this example.

As will be appreciated, the effective use of this graph requires that CBR figures are available. However, simpler tests can be conducted in the trial holes which will give the necessary information, provided that conclusions always err towards safety.

Solid rock (CBR 100%) is not difficult to recognize, provided it is clearly not friable (that is, has a consistency similar to sandstones, gritstones, limestones or solid chalk).

Gravels are also easily recognized and if a pick is required for their excavation then they can be considered to have a CBR of 40%.

Sands and sandy gravels (CBR 10%) whilst still very difficult to excavate with a spade, will allow a 50 mm (2″) square peg to be driven in up to 150 mm (6″) before the peg starts to split badly.

Silty and sandy clays (CBR 5%) can be excavated with a spade but lumps of the excavated material cannot be moulded in the fingers for there is insufficient clay to hold the mould. A 50 mm (2″) square peg can easily be driven in.

Light clays (CBR 4%) are more difficult to excavate with a spade but lumps of the material can be moulded in the fingers.

Heavy clays (CBR 3%) need a pick or mattock for their excavation and can only be moulded in the fingers with considerable pressure.

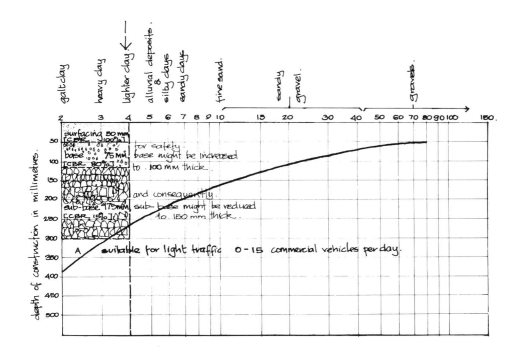

Subsoil materials other than those described above are best considered to be very unstable and expert advice should be sought.

Finally on the use of this graph; when the availability of particular sub-base and base materials is not known it can be used to give information for a performance specification. That is, taking the example used above, the road might be described as to consist of a 150 mm (6″) consolidated sub-base with a CBR of 15%, 100 mm (4″) consolidated base with a CBR of 80%, and a 50 mm (2″) surfacing of the kind required.

1.6 *This example illustrates how the CBR curves shown in Fig. 1.5 can be employed.*

1.7.3 Subgrade and formation

See section 1.6.5 (v) for the preparation of subgrade and formation.

1.7.4 Sub-base and base

All the materials should be very thoroughly compacted to beyond the point where there could be further compaction when the road is in use. Moreover they should be compacted in layers when the individual depth of sub-base or base exceeds 150–225 mm (6–9″). The way in which forces are spread as they pass down through a road structure has already been described. This will be the case as much when the road is being built as when it is in use. If attempts are made to compact a layer of material with a loose depth of more than 225 mm

(9″) (less if only a light roller is to be used), then compaction will be achieved only in the top part of that layer and the sub-base or base as a whole will not have the necessary strength. It is always preferable to use a heavy roller but if that is not possible materials should be compacted in layers with a loose depth not exceeding 100 mm (4″).

When on weak subgrades it will be cheaper to increase the depth of sub-base rather than the base, both because the material could well be cheaper and, because it does not need such high strength, it can be consolidated in thicker layers.

Sub-base materials usually have a CBR of between 15 and 20% when compacted. Clinker, quarry waste, low quality gravel, sand and burnt colliery shale, as well as some demolition materials, are all suitable. Base materials should be capable of achieving a CBR of 80% or more after compaction. They must be stable when wet and unaffected by frost. BS812 (parts 1, 2 and 3:1975) covers the taking of samples. On larger projects samples should be submitted for testing and approval prior to their use. Suitable materials include crushed stone, crushed blast furnace slag (covered by BS1047: 1983), and hogging (a naturally occurring graded mix of gravel, sand and clay which may be locally available but which may need to be sealed with a hot tar or bitumen spray prior to surfacing).

A rather obvious point (but one which has been known to be overlooked), must be made. The maximum size of the individual pieces within a sub-base or base material must be specified – compacting material with a potential CBR of 80% to a layer of whatever thickness would be somewhat difficult if it contained pieces that were larger than that thickness. A convenient rule of thumb suggests that a maximum size of 50% of the compacted thickness of the layer will prove satisfactory.

All of the materials described above will in themselves give roads which are adequate, in certain circumstances, without further treatment. Their appearance would at best be **casual** and mostly distinctly **utilitarian**, although that character would be modified by both the line of the road and the form of edging it was given. They would have a short life if heavily used and most would tend to be dusty – in both cases due to the absence of a sealed surface keeping dust in and water out. But, if vehicular traffic is very light and/or very occasional (and certainly if traffic is pedestrian only) and surface colour and texture is appropriate, then such pavements can be valuable.

1.7.5 Bitumen and tarmacadam surfacings

In order to avoid confusion short notes on the meaning and function of various parts of the surfacing are given here:

(i) *Surfacing*. The material which protects the structure from surface water and provides a suitable wearing surface to the pavement. It can be laid in one or two courses.

(ii)   *Single course work.* Surfacing laid in one course and functioning as both base and wearing courses.

(iii)   *Two course work.* Consists of base course (not to be confused with the road base) which protects the base from the traffic load, and provides a suitable surface for the wearing course which provides a waterproof and acceptably smooth non-skid surface for traffic. It will retain these qualities and continue to protect the base course, until worn away (whilst single course work will start to deteriorate in these respects from the outset).

(iv)   *Surface dressing (not to be confused with surfacing).* Small aggregate and binder, mostly used to renew the wearing course of an old road and sometimes used on new work when anticipated wear is low. Further, a surface dressing might be considered with new work in order to give the fine textured effect that can do so much to reduce the scale of the kinds of surfaces being considered here.

(v)   *Macadam.* The stones which in the original roads by Macadam held together by their mechanical bond, but which are now generally bound together with a binder.

(vi)   *Binder.* The viscous material which binds the aggregate together. It may be tar, bitumen, or a proprietary mix of tar, bitumen and (possibly) synthetics.

(vii)   *Tar.* A product of the distillation of coal. Crude tars are refined to produce road tars and are supplied in bulk to firms producing tarmacadams. The cost of tarmacadams depends to a significant extent upon the distance between such a plant and the site.

(viii)   *Bitumen.* A naturally occurring mineral pitch or a product of the distillation of oil. Generally it is a little more durable than tar when exposed to the atmosphere.

(ix)   *Asphalt.* A mixture of mineral aggregates of various sizes from powder to coarse sand, with bitumen. Naturally occurring in some parts of the world but principally available as proprietary mixes sold under many trade names. It can be used hot, giving a very durable surface even with heavy traffic loads, but with high initial costs, or cold, giving lower initial costs but retaining high durability for lightly trafficked roads and footpaths.

(x)   *Aggregates.* Crushed rock, or rock-like industrial by-products or demolition materials, or gravel. Gravel is not recommended for highly stressed roads as its roundness makes for poor adhesion and mechanical bond.

1.7.6 Factors influencing choice of surfacing

The surfacing must be both waterproof, durable and non-skid, and aesthetically satisfactory in its context. Unfortunately, this combination of qualities is not simple to achieve. The practically better surfacings tend to have little visible texture and a consistent black colour. More open textures and the use of uncoated aggregates (when native colour can still be seen) offer more to engage the eye but give a technically less satisfactory protection to the structure below. The use of uncoated chippings in a surface dressing is a way of getting the best of both worlds.

(i) *Appearance.* The designer must be interested in appearance as well as durability. Even the narrowest of roads cover a large area of floor-scape and a car park or delivery area much more. Although the character of the road or drive is significantly determined by both its line and its edging details, both colour and surface texture also play an important part. Smooth black surfaces immediately suggest intense urbanization and heavy traffic. As they are generally seen (or, rather, largely unseen) with a throng of buses, lorries and cars, lack of texture and colour variation is probably unimportant. But their black reflectivity in the deserted city street (in the more romantic images, it has just stopped raining) does seem effectively to match character and context. The unbroken surfaces, simply contained, do augment the scale of the city as a whole. But in the country where it should be the broken textures of grass and leaves which establish scale and character, those qualities can seem dull if not dead.

(ii) *Durability.* In open-textured surfaces the binder is exposed to both oxygen and ultraviolet light. Rapid hardening of the binder will cause premature disintegration of the surface. Moreover, open textures have less integral strength and the road will also be more vulnerable to both wear and the effects of water penetration. Specialist advice is always likely to recommend more closely textured surfaces. With heavily trafficked roads, the wear in itself can compensate for initial appearance. The hardened binder is confined to the very surface of the surfacing. Traffic will quickly wear away that hardened material giving colour variation (if that is in the aggregates used) and apparent texture through that colour variation. This will also give a good skid resistance. But only when traffic volumes are high. So, the matter must be judged in the light of traffic loads. Heavy traffic needs durable roads and such surfacing is likely to be in scale and character with such contexts. But with lightly trafficked roads, colour and texture may well have to be designed in, and this can be done with

Technical alternatives must also be considered in
terms of their effect on scale.

some confidence knowing that absolute durability is of lower importance.

(iii) *Non-skid surfaces*. The degree of skid resistance required is clearly conditioned by traffic speed, but everyone should be interested to some degree. Nearly all roads have good resistance to skidding when dry. The principal exception, unsealed roads with a loose top surface, are more frequently used at low speeds and so a problem does not arise (the antics of rally drivers illustrate the point). Skid resistance, when the road is wet, is given because the tread pattern on the moving tyre and the texture of the road, in combination, throw off the water which otherwise would remain as a thin film between tyre and road. It is only in the extreme cases of heavy braking or acceleration that skidding occurs due to insufficient friction between tyre and road, and even then, in large part, the skid results from the film of rubber that is rapidly worn from the tyre. Some aggregates are not suitable for high speed roads for they polish smooth and road texture becomes minimal. As we have seen, surfacings with open texture are not suitable for high speed roads, even though their initial skid resistance is high, because they wear too rapidly. The road texture required, when speeds are high, is relatively fine, with a high proportion of smaller aggregates which are less likely to alter over time.

Clearly matters such as the minimum radii appropriate for roads designed for different traffic speeds, is beyond the scope of this book. But what can be said is that should a designer sense that questions of that kind are being generated by design work in hand, it should be taken as a clear signal that specialist advice is needed.

(iv) *Prior use of sub-base*. Contractors will often lay the sub-base as a very early operation, so as to allow free access to and within the site, both for staff and vehicles. The surfacing is added as a very late operation, after the time at which it might be damaged by heavy vehicles. That is good practice, but care should be taken to ensure that the base has not been damaged during the interim period by traffic and/or water penetration. If it has it must be repaired. There may be advantage in sealing the base with a bitumen spray and sand dusting, so as to keep water out of the road, provided that there is provision for drainage.

In some circumstances it could be useful to deliberately include a sub-base in the design of a road or drive which strictly speaking did not need it, so that it is available for use by construction traffic. The presence of this sub-base could affect decisions about surfacing.

(v) *Single- and two-course surfacing*. With lightly trafficked roads this is likely to depend entirely on the thickness of surfacing required (for the provision of an easily repairable wearing course is less likely to be a requirement). The illustrative calculation in section 1.7.2

shows how surfacing thickness can vary with loadings and base material strength. Further, the thickness of surfacing might be conditioned by the texture that is to be achieved. Coarse texture is given by larger aggregates and, as with sub-base and base materials, there is a minimum thickness that can be made with aggregates of a particular size – the larger the aggregate the thicker must be the layer. In this case the convenient rule of thumb is maximum aggregate size not greater than 66% of layer thickness (for example, a 38 mm (1.5″) course should have no aggregate larger than 25 mm (1″).

A surfacing course over 100 mm (4″) is not practical; attempts to roll in one part are likely to result in an upwards heave in adjacent parts, rather than consolidation. Further, surfacing courses of over 100 mm (4″) would be very expensive and it would almost certainly be cheaper to use stronger materials for the base, should they be available. So, if necessary or desired surfacing thickness is down towards the thinner end of this range (say below 38–63 mm (1.5″–2.5″) one-course work is the only likely option. If it is over 100 mm (4″) two courses must be used. One-course work is cheaper initially but might involve higher maintenance costs. Broadly speaking a two-course surface might last nearly twice as long as would one course on an otherwise identical road, before needing repair.

Surfacing courses of less than 63 mm (2.5″) are not recommended for roads carrying any significant amount of traffic, other than as a repair to an existing road. However, surfacings at a minimum 38 mm (1.5″) have shown themselves to be adequate for pedestrian areas with very occasional light vehicle use, private drives, etc. and less can be used for pedestrian-only access.

Finally in this respect, total thickness of surfacing and the number of courses it should comprise, are also conditioned by the regularity of the surface of the base. Bases using larger material will be less regular than those using smaller. Such regularity is measured by placing a 3 m (10′0″) straight-edge on the base surface (at right angles to any camber or fall) and measuring the size of the largest gap found beneath it. If this is over 13 mm (0.5″) a surfacing up to 100 mm (4″) thick should be used in two courses; if below 13 mm (0.5″) a surfacing between 50–100 mm (2–4″) should be used in one or two courses; if below 10 mm (⅜″) a surfacing below 50 mm (2″) might be used in one course; all providing that the base is strong enough for the thickness of surfacing used, and that the size of aggregates employed are not too large for course thickness.

To summarize: a very lightly trafficked road might be given a number of different surfacings and surface thicknesses, of which three examples follow.

(a) A 30–38 mm (1¼ – 1½″) wearing course, maximum aggregate size 20 mm (¾″), with a 38–50 mm (1½ – 2″) base course, maximum aggregate size 25 mm (1″).

(b) A single course 63 mm (2½″) thick, maximum aggregate size 38 mm (1½″).

(c) A 20 mm (¾″) wearing course, maximum aggregate size 10 mm (⅜″), with a 60 mm (2¼″) base course, maximum aggregate size 38 mm (1½″).

Three examples of surfacings for pedestrian-only areas might be:

(d) A 13 mm (½″) wearing course, maximum aggregate size 6–9 mm (¼ – ⅜″), with a 38–50 mm (1½ – 2″) base course, maximum aggregate size 25 mm (1″) – tarmac.
(e) A single course 25 mm thick, maximum aggregate size 6–9 mm (¼ – ⅜″) – bitmac.
(f) A single course 20 mm thick – cold asphalt.

With all these examples, thicknesses are after compaction. All bar the very thinnest might also be surface-dressed with chippings. It should be clear that all must be on an appropriately designed and built base. It should be equally clear that these examples are alternatives for different circumstances.

(vi) Choice of binder. The examples above make reference to tarmac (tarmacadam), bitmac (bitumen macadam) and cold asphalt. The essential difference between these is the material used to bind the aggregate. (The examples might also have included hot rolled asphalt.) Tar is generally cheaper as a material, when available and when transport costs are equal. But tar is generally less durable as it tends to harden quicker when exposed. A tar-bound road might, say, not need resurfacing for four or five years, whilst one which was bitumen bound might last six or seven years under similar circumstances. Tar is also more susceptible to extremes of temperature. (It is both softer when hot and more brittle when cold.) It is, therefore, quite common for tarmac to be specified for a base-course and bitmac for the wearing course, of a two-course surface. If, however, the road is to be surface-dressed then tarmac might be used throughout, for it will have that surface dressing to protect it. But, tar surfacings are not softened by oil droppings, as bitumen tends to be, and so might be more appropriate for garage forecourts, parking areas, etc.

Asphalt, somewhat paradoxically, might be considered most appropriate at the extremes of the pavement market. This material is very durable and its smaller aggregates give excellent non-slip characteristics. Its relatively high costs can, then, be entirely justified as the wearing course of high-class, heavily trafficked roads. At the other extreme its higher costs might be justified for pedestrian-only areas when its strength to thickness ratio can be exploited to limit overall construction depths (and thus, costs).

Pigments can be added with the binder to radically modify the colour of the surfacing, but red (starting deep pink and weathering to reddish-brown) is the only one commonly available, with green being occasionally available to order.

Proprietary macadam-type mixes are available for surfacings (really wearing courses only) where the normal black bitumen binder is replaced with a clear synthetic resin. The clear binder leaves the aggregate open to view and so is invariably used with very decorative aggregates. Their costs are generally comparatively high.

(vii) Choice of aggregates. Stones with medium-textured surfaces give the best adhesion, particularly when the form of the individual pieces is irregular. Stones with, or capable of acquiring, smooth glassy surfaces, and/or which are crystalline in nature, should be avoided. Similarly gravels (which are likely to have an essentially ovoid form due to their geological history) are less preferable to crushed stone. Choice of aggregate will depend largely on that which is locally available except, perhaps, in the case of a surface dressing when an imported material might be used for its aesthetic qualities. BS 63:pt2: 1981 covers grades of crushed stone and slag. Grades, in this case, refers specifically to maximum and minimum aggregate sizes, together with the proportions of the various sizes between those extremes. Appropriately graded aggregates are important if mechanical bond is to be achieved, as smaller stones lodge in the interstices between the larger ones during compaction. Aggregates coated in clear synthetic resins are proprietarily available. The aggregates given this special treatment are all highly decorative. The resins protect the aggregates from dirt and weathering and so the highly decorative finish is retained. Their relatively high cost makes them suitable for use only in thin layers, that is for the wearing course of two-course surfacings. Alternatively, they can be scattered over and rolled into the surfacing as a final operation. Care must be taken with their use for they might create practical difficulties in terms of skid resistance, and their decorative qualities might make an area of any size seem distinctly garish. For all these different reasons they are probably most appropriate for footpaths or very light traffic only, in contexts that are essentially bright and breezy.

1.7.7 Compaction of surfacing

After laying, the surfacing must be rolled to achieve necessary compaction and strength. There is a direct relationship between the weight of roller used and strength achieved through compaction, given similar specifications and thicknesses. Weights of rollers likely to give the best results in different circumstances are:

(a) Regularly trafficked roads – 6–8 tonnes (6–8 tons)
(b) Lightly trafficked roads and heavily used pedestrian areas – 1–3 tonnes (1–3 tons)
(c) Pedestrian-only footpaths and the like – 0.5 tonnes (10 cwt)

Particular care must be taken to ensure that surfacings are as well compacted in the awkward places as they are elsewhere. These will be at edgings and around inspection chamber covers, gully gratings, etc. These are likely to be best treated with pedestrian operated vibrating-plate machines, which are generally rated in terms of their roller weight equivalence. Where awkward places abound and/or the equivalence of a very heavy roller is not available, consideration might be given to using thinner courses compacted solely by a pedestrian-operated machine. That might also be the case where the nature of the site itself would make access for a large machine difficult, or where the total area to be surfaced is small.

1.7.8 Surface dressing

A surface dressing can be applied to an old road in need of a new waterproof surface, or to a new road when there is an immediate need for a particular colour and/or texture without waiting for wear to expose the potential in the road itself.

Remember that with a narrow road, where wheel tracks are all on virtually the same line, the aggregate in the surfacing may never be exposed by wear except on those lines. Surface dressing might be useful in such a case, but even then a textural difference between the tracked and untracked strips will soon emerge. But this in itself can have aesthetic advantage, particularly when informal character is desired.

Remember also that the chippings in a surface dressing are not thoroughly incorporated into the road surface. With more heavily trafficked roads any loose material is thrown off the surface (hopefully not through a windscreen) in a matter of days, but when densities and speeds are low that may not happen for some time, and with narrow roads it may never happen in the central strips. Roads covered with loose material have very poor non-skid properties.

The maximum size recommended for the aggregate in a surface dressing is 13 mm (½″). A wide choice of colour is available. Whilst this will be governed to an extent by local availability, relatively little material is needed for a dressing and so transport costs, for once, need not be an overriding consideration.

Both tar and bitumen can be used as a binding spray with choice being determined by the same considerations as discussed earlier. Tar is most frequently employed, for financial reasons. Little of the binder is exposed to the atmosphere and so rapid deterioration should not be a problem. However, weather conditions can affect the decision. Should there be a risk of heavy rain or low temperatures during the day or two after laying, bitumen binder would be a better choice. Alternatively (or additionally) adhesives are available which can be applied to (or with) the binder film before the chippings are spread. An alternative form of surface treatment, bitumen slurry coating, might be considered. This consists of a bitumen emulsion and fine aggregate mixture which is applied to the surfacing either by brush or squeegee to a thickness of approximately 13 mm (½″). The aggregate used can be coated with a clear synthetic resin.

A final form of surface treatment that could be appropriate spreads and rolls fine aggregates (3–6 mm (⅛″ – ¼″)) directly into the surfacing as a final operation. This is only possible when the nature of the binder and the size and proportion of aggregates in the binder allow. It is not, then, a suitable treatment for more stony bituminous mixes or high stone-content hot-rolled asphalts.

1.8   COLD ASPHALT SURFACING

1.8.1   Uses

The particular advantage of cold asphalt is the very fact that it does not have to be heated on site, making site works a somewhat simpler operation. This, together with the fact that it needs only relatively light and manoeuvrable machines to achieve necessary compaction (it being laid in thinner courses than some other surfacings), means it can be used in relatively small and awkward areas. Whilst the material has a value in the design of lightly trafficked roads, it is more frequently used for footpaths, school playgrounds, etc. As a material it is rather more expensive than tarmac or bitmac, but these costs might well be offset by on-site convenience and low maintenance costs. A cold asphalt surface should not need attention for up to ten years.

Suppliers should be informed of the purpose for which the material is to be used, and whether it is to be used immediately or put into storage. The contractor should be directed in the specification to ensure that this is done.

1.8.2   The material

Cold asphalt is a mixture of bitumen and crushed igneous rock, limestone or slag with a maximum aggregate size of 6 mm (¼″). The aggregate should be clean and hard and of a kind that will compact solidly. The material also contains a filler which is, essentially, the very fine material in the graded aggregate, but other constituents such as hydrated lime may be used. The bitumen binder is specially prepared for this purpose.

1.8.3   Construction

Cold asphalt is best used as a wearing course on a tarmacadam base course. But in lightly loaded circumstances it can be employed as the whole of the surfacing. The surface of the base course (or base) should be such that there is no gap larger than 9 mm (⅜″) below a 3 m (10′0″) straight-edge.

When used over a tarmac base course that base course itself must be specified as to be appropriate for cold asphalt wearing course. (More precisely, the matrix content of the tarmac base course must be towards the lower limit provided by BS4987:1973.)

If the tarmac base course is not freshly laid, or if the cold asphalt is to be laid directly on the base, it should be given a tack coat of bituminous emulsion sprayed on at a rate of 2–3 m^2/litre (10–16 sq. yd/gal.) Particular care must be taken to ensure that this spray is applied at any daywork or other butt joints in the base course. The cold asphalt material can be spread either by hand or by machine.

1.8.4 Thickness

As always, this will depend upon the loads that will come on it in use and the strength of the base course or base beneath. A compacted thickness of 9–19 mm (⅜″ – ¾″) for a wearing course is likely to be satisfactory in pedestrian-only areas, when compacted with a 0.75–2.5 tonne (¾ – 2½ ton) roller, or a vibrating-plate equivalent.

A compacted thickness of 13–19 mm (½″ – ¾″) for a wearing course is likely to be sufficient for very lightly trafficked roads, but then it would be best to compact with a 5 tonne (5 ton) roller or its equivalent. A compacted thickness of 19–25 mm (¾″ – 1″) would be recommended when cold asphalt is used for the whole of the surfacing to a base, with a heavier roller being used if light vehicular traffic is anticipated.

Paradoxically, correct compacting is most important in pedestrian-only areas. Foot traffic can be hard on this surfacing, particularly when it is that of a number of playing children, for the oblique forces as they run and turn can tend to 'scuff off' the surface, but such traffic does little or nothing to help with compaction. Nor does it help to leave the surface for a period after laying, before opening it to the children. Strength is due solely to continued compaction and nothing else.

When used as a sealing coat to tarmac or bitmac roads, the surface should be brushed over with cold asphalt sufficient to fill voids (a coverage of about 200 m^2/tonne (175 sq. yd/ton), before rolling the sealing coat at about 80 m^2/tonne (75 sq. yd/ton) giving a compacted thickness of approximately 6 mm (¼″).

1.8.5 Finishes

Two types of finishes are described:

(a) Sandpaper. The self-finish of the material; non-skid, fine-textured but relatively smooth in appearance.
(b) Roughened. Either coated or uncoated chippings can be rolled in to modify surface texture. If uncoated chippings are used, colour can be modified as well.

A sandpaper finish requires no further treatment once laying operations have been properly completed.

A roughened finish can be given by rolling chippings into such a surface. Where adhesion between such chippings and surface is critical, they must be appropriately pre-coated for this purpose. This will modify the texture of the surface but not its colour. But on footpaths and other pedestrian-only surfaces, where the forces that will attempt to pluck the chippings away are lower, uncoated chippings can be used to effect.

Chippings, whether coated or uncoated, should be rolled into the cold asphalt surfacing immediately after laying. They are generally single sized (that is, not graded) between 13 and 25 mm (½" – 1") when coated and between 3 and 6 mm (⅛" – ¼") when uncoated. The weight of the roller used, as elsewhere, will increase with anticipated traffic load from, say 0.75 tonne (15 cwt) for a footpath, to 2.5 tonne (2½ ton) for a lightly trafficked road.

Rates of application should be: 13 mm (½") chippings – 90 to 145 m²/tonne (100–160 sq. yd/ton); 19 mm (¾") chippings – 75 to 110 m²/tonne (85–120 sq. yd/ton); 25 mm (1") chippings – 55 to 90 m²/tonne (60–100 sq. yd/ton); all these being pre-coated for light roads. Decorative chippings of 3 to 6 mm (⅛" – ¼"), that are coated with a clear synthetic resin, can also be used for very colourful and very active pedestrian areas.

The use of chippings to give a roughened surface to cold-rolled asphalt can interrupt surface water drainage. There can be an advantage in leaving any channels or gullies free of chippings when it is necessary to get the rainwater off at a reasonable rate. But remember that there could be aesthetic disadvantages associated with this. Conclusions as to where such channels should be, must be informed by both visual and practical considerations – as discussed at some length in the introduction to this section.

1.9 CHIPPINGS SEALED WITH COLD BITUMINOUS EMULSION

1.9.1 Uses

Where loose chippings are likely to be impractical, but where a comparatively cheap and informal surface is needed, bituminous emulsion binders can be useful. As with other such areas, thorough rolling is necessary, so the area must be accessible and have appropriate shape. However, if smaller aggregates are used in thinner courses, a pedestrian-operated vibrating-plate machine can be used, and as both the chippings and the binder can be spread by hand, small or otherwise awkward areas can easily be treated in this way. But such a surface would be inappropriate for anything other than the lightest traffic loads. In some cases only a part of the area is sealed (at entrances for example, or elsewhere that traffic will be turning, starting, stopping) with the remainder being finished in the same stone but without the bituminous emulsion spray.

1.9.2 Types

There are a number of proprietary cold bituminous emulsions. The manufacturer's instructions should always be most carefully followed. Some manufacturers will provide a list of contractors with experience with their product(s). All will gladly advise. There are no really significant differences between their products and so choice will be determined by availability, experienced contractors and their preferences, and maybe cost. All are eventually black; it is the colour and texture of the aggregates that determine appearance.

1.9.3 Construction

Any thoroughly compacted chippings or gravel surface can be sealed with bituminous emulsion, but if sealing an existing surface it must be remembered that the rainwater which previously percolated through that surface will now run off, and so it is likely that some provision for surface water drainage will have to be made. Consideration should also be given to the fact that the roots of adjacent trees and shrubs might be sealed from the air. If that extends over more than ¼ to ⅓ of the root area, the plant will almost certainly die, particularly if it is quite old and so unable to adjust to the change of environmental circumstance. If worried, areas close to existing shrubs and trees should, where possible, be left unsealed.

Work should be carried out in reasonably dry weather. Irregularities (pot-holes or whatever) in an existing surface should be made good, and a well consolidated base (be that old or new) prepared and shaped to falls. All weeds and moss should be grubbed up, not simply killed, for their later decomposition would affect long-term stability. There would be value in treating the surface with a persistent weedkiller to avoid later problems, but adjacency of plants that are cherished might make that impossible. The work can be done on a do-it-yourself basis, carefully following the manufacturer's instructions and using an old watering can (with a special spreader supplied with the material when sold for this market) and a stiff garden broom.

1.9.4 New drives

On a well consolidated base with a thickness appropriate for anticipated loads, a 50 mm (2″) layer of 38 mm (1½″) clean angular stone (rounded stone cannot be properly consolidated) is spread and consolidated with a 6–8 tonne (6–8 ton) roller or its equivalent. Cold bituminous emulsion is then spread at a rate of 3.5 – 4.5 litre/m²

(¾ – 1 gal./sq. yd) and immediately covered with 13 to 19 mm (½″ – ¾″) clean limestone, granite or other hard chippings, sufficient to fill the interstices in the surface of the base (say 80 m²/tonne (90 sq. yd/ton)). The surface should be rolled immediately, and rolled again the following day. The area should then be opened for use and when thoroughly dry (perhaps 10 to 14 days) it should be swept and a sealing coat of emulsion applied at a rate of 1.5 litre/m² (⅓ gal./sq. yd). This in turn should immediately be given a dressing of 6–9 mm (¼″ – ⅜″) chippings at about 115 m²/tonne (120 sq. yd/ton).

1.9.5   New paths

On an appropriately designed and constructed base, spread sand or clean ashes sufficient to fill surface interstices, water well and roll. On this spread a layer of clean 19 mm (¾″) chippings or sharp gravel at 25 m²/tonne (30 sq. yd/ton). Again water well and consolidate. Bituminous emulsion should be applied at a rate of 2.5 to 3.0 litre/m² (½ – ¾ gal./sq. yd) and evenly covered with 6 mm (¼″) chippings at a rate of 100 m²/tonne (120 sq. yd/ton), and rolled. Two days later (this is a minimum) when the emulsion has set, a second dressing of 6 mm (¼″) chippings should be spread at a rate of 135 m²/tonne (150 sq. yd/ton) and well rolled.

1.9.6   Existing drives and paths

Loosen the surface of the existing drive or path to a depth of 25–38 mm (1″ – 1½″) and bring to required levels. If very dusty, water well and roll. If the base has a high proportion of fine material, spread a layer of 19 mm (¾″) chippings at a rate of 35 m²/tonne (40 sq. yd/ton). Spread bituminous emulsion at a rate of 2 to 3 litre/m² (½ – ¾ gal./sq. yd). Immediately spread 6 to 9 mm (¼″ – ⅜″) chippings at a rate of 90–135 m²/tonne (100–150 sq. yd/ton), roll, and leave for up to 7 days. Seal with emulsion at a rate of 1.5 to 2.0 litre/m² (⅓ – ½ gal./sq. yd), and dress with 6 to 9 mm (¼ – ⅜″) chippings at a rate of 115 to 135 mm²/tonne (130–150 sq. yd/ton), and roll. Roll again the next day if surface is to take light vehicular traffic. With all quantities, sizes, rates, etc. given above, the higher figures apply for lightly trafficked surfaces whilst the lower are satisfactory for pedestrian-only areas.

1.10   CHIPPINGS SEALED IN A SYNTHETIC BINDER

1.10.1   The material

A number of companies will supply a clear synthetic resin binder together with (usually) brightly coloured aggregates of a consistent size – between 3 and 6 mm (⅛″ – ¼″). The costs of such materials are relatively high and so they are used solely as a wearing course that is only a little thicker than the size of the aggregate. (With this material there is a much higher proportion of binder to aggregate than in other cases and that binder has some strength, and so the general rule of thumb about maximum aggregate size and minimum thickness of layer does not apply.) Being clear, the binder exposes the aggregate to view whilst keeping it free of dirt and the effects of the weather.

Similar aggregates are also available individually coated in clear synthetic resin. These also keep the aggregates clean and unweathered. They can be scattered over a bitmac or tarmac surfacing and rolled in as a final operation. Adhesion between them and the other constituents of the surfacing is poor, and so that course would be weakened. But when that weakening can be accepted, such chippings can effectively relieve the appearance of 'black-top'.

As with all proprietary materials, the manufacturer's instructions give the only reliable guide to their use.

1.10.2 Construction

Construction below this type of finish should be exactly as with surfacings with other binders – all dependent upon anticipated loads, base course and base strengths, and subgrade capacities. Indeed, it would be best to consider it as a surface dressing rather than as a surfacing and ensure that the construction is designed to be functionally satisfactory without it.

1.10.3 Characteristics

Non-skid characteristics are rather poor; the surface gives little physical texture even though its apparent texture is high. This, together with the boldness of the material which could make an extensive area somewhat daunting, and its price, makes it unsuitable for roads.

1.11 UNSEALED CHIPPINGS OR GRAVEL

1.11.1 Uses

Loose gravel or chippings have an essential informality, given partly because the material clearly could not be used intensively, partly because its nature encourages its use in irregular shapes, and partly because its distinct but consistent texture holds or at least slows the eye, inducing a general calm. It can often be used as an alternative to grass, in circumstances where grass would not survive. It is particularly useful where trees are growing, as a rigid line at their trunks can be avoided and, because the surface is not completely sealed, air is allowed to the roots (but the better the rolling the less this applies). Car parks amongst trees, for example, can be very successfully surfaced in gravel. It is also very suitable for private drives where little traffic is expected and where the surface is unlikely to be disturbed by violent acceleration or braking.

1.11.2 Types

Graded quarry chippings essentially similar to the aggregates used in bound pavements, are most satisfactory when traffic loads are likely to be towards the higher end of the lightly trafficked category. Less carefully graded, or single sized, chippings might well be a little cheaper, would be very satisfactory for very lightly trafficked areas, and are more likely to have the textural qualities described above. Naturally occurring gravels will be more rounded in form and so will achieve less mechanical bond when compacted, making them less useful with any intensity of traffic. However, most have more colour variation than stone chippings and so may offer a visually more satisfactory surface. In either case, local sources will be cheaper than imported materials, and it is probable that local materials used in a local context will better exhibit the informal characteristics considered earlier.

1.11.3 Construction

Many specifications exist for gravelled surfaces – almost as many as there are individual gravelled surfaces. With all materials it is imperative that the designer understands the principles which guide their use, and responds to the specifics of each case within those principles. This is particularly the case with gravelled surfaces where there will be significant differences between the products of different localities. As an example, a material with a higher proportion of fines will compact better and so can be used in thinner layers to achieve a certain strength. But at the same time, those fines will more effectively seal the surface and so surface water run-off and drainage will have to be more carefully considered.

For good results, particularly when motor traffic is anticipated, heavy rolling is essential at all stages of the job. Otherwise the surface will soon become very loose and ruts will start to appear. This remains the case when other wheeled vehicles, such as cycles, prams or wheelchairs, will use the area. Whilst these impose lower loads, and so compaction is not necessary in that respect, they will have great difficulty if the surface is loose and yielding. Watering during rolling is essential in warm weather.

A typical specification for a lightly trafficked drive might be: on an appropriately prepared formation lay 150–200 mm (6″–8″) unconsolidated layer of 100 mm (4″) base material and consolidate with an 8–10 tonne (8–10 ton) roller, watering well if the material is dry. Surface with a 50 mm (2″) unconsolidated layer of 25–38 mm (1″ – 1½″) local stone chippings and water and roll as above. Finish with 6–9 mm (¼″ – ⅜″) local stone chippings containing sufficient fines to act as a binder, in a 25 mm (1″) unconsolidated layer, water and roll.

1.12 SOIL–CEMENT BASES

There was some debate between the authors before it was decided that this section of the first edition of this book should be retained. The technique it covers was developed (particularly) by military engineers during the early 1940s who at that time had a pressing need rapidly to construct roads and airstrips in circumstances which encouraged the maximum use of locally available materials. In those cases, locally available was very narrowly defined to mean the subsoils immediately below the line of the proposed road or airstrip. The technique continued to have some popularity through the late '40s and '50s – indeed was promoted for certain uses such as on farms. This period of popularity was, one suspects, because many of those military engineers had by then returned to civilian employment and were keen to beat that particular sword into a ploughshare. Whilst it is a very economic form of construction when offered an appropriate subsoil, it has to be said that the design and construction of such bases is a somewhat hit-or-miss process unless very expert advice and very specialized machines are available. The economics of civilian life do not seem to have encouraged the retention of such specializations, either intellectual or mechanical.

The brief description of this technique given below is, then, offered primarily for historical interest, although it might guide somebody wishing to undertake do-it-yourself experimentation. (It should be noted that the above statement has been written in the British context where geological circumstances can change rapidly over very short distances. In other parts of the world where extensive lengths of new road could be built with little if any variation in subsoils, cement-stabilized subsoil bases can have real value.)

1.12.1 The technique

Light sandy soils are the most suitable for this technique. Heavy clays are unsuitable as are any with an organic content. Other soils might be adjusted by the incorporation of granular materials but the necessity of doing that could well negate economic advantages. High concentrations of agricultural chemicals, or any other forms of pollution, could adversely affect the setting cement. Soil tests must be undertaken both to establish that subsoils are appropriate and to determine the quantities of cement and water, and the degree of compaction, that will be required. The intervals at which samples are taken has to be judged on the results found from the first of such tests. The greater the variety the more frequent the tests.

Soil–cement is a thoroughly compacted mixture of pulverized soil and cement with water. The object is to strengthen the soil and make it resistant to water so as to form a satisfactory pavement base. French drains may have to be constructed and these should be located to allow the stabilized base to extend at least 300 mm (1'0") beyond finished verge.

The soil is pulverized to a fine tilth with a rotary tiller. The depth of this pulverization will depend upon anticipated loads and strengths. 75 mm (3") has been quoted for light sandy soils where the pavement is to be used by light traffic only. It may be necessary at this stage to add water if the soil is dry. Cement powder is then spread either mechanically or by hand and the soil and cement thoroughly mixed with a tiller and compacted by roller. The area must then be left for at least 7 days to allow the setting cement to gain strength. Evaporation of necessary water must be avoided either by covering with polythene film or by treating with a bituminous emulsion spray.

Soil–cement bases have a high load-bearing capacity but they have little resistance to abrasion. They must, then, be surfaced, the minimum requirement being a double surface dressing of bitumen or tar with rolled chippings.

Specialized machines, sometimes known as a road train, can undertake all of this work in one pass; pulverizing, mixing, watering and compacting with a series of linked vehicles. They can progress at rates of between 1 and 4 m/min (3–12'/min). But there needs to be many miles of road to construct if they are to be economic.

This form of construction is very cheap both to build and to maintain – philosophically, it seems a pity that it has gone out of favour. Perhaps increased interest in 'alternative technologies' will reverse that trend?

1.13   *IN SITU* CONCRETE PAVEMENTS

1.13.1   Scope

This section is concerned with concrete roads, pavements with concrete surfaces, car parking areas for light or very light traffic loads, pedestrian areas to which vehicles might occasionally be given access, and purely pedestrian areas. The information which is given about roads for high traffic densities is included solely to help establish technical context, and would undoubtedly be found to be insufficient in itself for the design of major roadworks.

1.13.2   Use

In situ concrete is one of the cheapest forms of paving. Unfortunately evidence abounds that suggests that cheap equates with nasty. There are no good reasons whatsoever for that to be so. The most common reasons for the poor appearance of some concrete pavements are:

(i) *Poor design.* The designer must understand the principles which govern the design of such structures and respond imaginatively to the specifics of any one case, within those principles. These principles include not only the theory of their design but also the ways in which that theory will be affected by site conditions before, during and after construction.

(ii) *Poor construction.* This would be the inevitable consequence of poor design, but the best of design efforts would be negated by the use of unsatisfactory materials, labour or workmanship.

(iii) *Unimaginative use of the various surface finishes available.* There are many of these and whilst some might add a little to the overall cost of the pavement, few would make the cost of the structure as a whole economically uncompetitive. The designer must not only know the range available but must also be able to decide which would be appropriate in particular contexts, and which it would be reasonable to expect to be within the scope of the kind of contractor that will be employed, and the circumstances in which he will have to work.

(iv) *Poor edge detail.* As with all pavings, *in situ* concrete drives and paths must be appropriately detailed at their edges. Physically, there is a concern that the edges are well supported to avoid subsidence and inevitable cracking. Visually, there will be a concern either to give the concrete surface an individual identity through expression of its edges, or to follow a strategy of integration by answering structural questions within the construction of the path itself.

Any of the following circumstances might suggest the use of an *in situ* concrete pavement.

(a) When a concrete finish is required. Exposed aggregate and other textured finishes to concrete can be considered for their positive qualities. Whilst these pavements have utility it would be wrong to consider them as purely utilitarian. Clearly when it is a concrete finish that is wanted, the structure below will itself have to be concrete.
(b) When more than one basic form of pavement is to be included in the same area. For example, bricks or flags might be intended for the pedestrian-only areas of a scheme and a decision taken to use the same materials to delineate a footpath across a parking area, to demark the parking spaces, and to treat the surfaces immediately surrounding trees. It could prove very difficult (if not impossible) to roll a flexible pavement in such circumstances whereas a concrete structure could overcome the difficulty with comparative ease.

(c) On poor soils. Concrete pavements are at their most economic on poor subgrades where the excavation costs for a flexible structure might prove prohibitive.

(d) Where the nature of locally available materials, or the proximity of a ready-mix or cement works, would show savings on transportation.

1.13.3 Properties of an *in situ* concrete pavement

Concrete as a material has some tensile strength, and it is technically easy to increase that considerably by the inclusion of steel reinforcement. Tensile strength gives a resistance to bending. Because a concrete slab bends much less than does a flexible structure, it does not spread the forces from an imposed load over the area immediately below that load, but over a much wider area. So, whilst the basic principle that it is the subgrade which supports the load applies equally to both flexible and rigid structures, with the latter the depth of construction needed to give adequate spread is less in any given circumstance. Sub-bases are generally not needed and in some cases a base may not be required either, whereas with flexible pavements the design of the base is the main structural consideration. Conversely, where a sub-base is required to make up levels (for example), much of the economy in an *in situ* concrete structure would be lost.

1.13.4 Design of a concrete slab

Before discussing the concrete itself, factors governing slab thickness will be briefly described.

(i) Soil. The strength of the subsoil below a concrete road has nothing like the same direct effect as it has with a flexible pavement. That being so scientific testing methods, such as CBR, are not required. If subgrade bearing capacities are low it is more likely that reinforcement will be introduced (or increased) rather than overall depth of construction being significantly affected.

(ii) Traffic loads. These are classified in the same ways as for flexible pavements.

(iii) Preparation of formation. Again this is the same as with flexible structures.

(iv) Base. If, having removed the topsoil and any subsoil with organic material (such as peat), the correct formation level has been reached and with a reasonable subgrade, then it is likely to be structurally unnecessary to have a base. However, it may prove difficult in practice to get an acceptable working surface on some subsoils and then a base would be useful. (Some clays, for example,

43

get very sticky and impossible to work on.) If there is any doubt prior to construction starting, it would be best to include a base in the specification, with that item being later omitted should it prove unnecessary.

A base is obviously necessary when the ground has to be made up to desired levels, and it may also be necessary to encourage drainage if the subsoil is susceptible to frost damage, as are chalk, silt and soft weathered limestones. A base should also be used: if there is reason to believe lenses of peat to be below the site, or if the subgrade is a very plastic clay, both of which being liable to non-uniform movement; on embankments over 1.2 m (4'0") high; and on subgrades where groundwater might rise to within 600 mm (2'0") of the formation.

When a base is necessary it should be essentially similar to those provided for a flexible structure although it is very likely that it can be a less deep layer. A cement-stabilized soil base might be considered, as might a lean concrete base (that is one with a proportion of cement to aggregates of between 1:15 and 1:20). If broken stone or clean broken demolition material is used, a reasonable quantity of smaller aggregates and fines must be present, both to ensure that a sufficient mechanical bond is achieved through compaction, and also to mini-mize the amount of relatively expensive concrete that is lost from the slab into the interstices within the base.

The base, after consolidation, should have at least the strength of the subgrade. 75–100 mm (3"–4") is a practical minimum although a 50 mm (2") run of smaller aggregate might be used where the base is not expected to contribute greatly to the strength of the pavement but is needed to improve working conditions.

1.13.5 Reinforcement

It should be made clear from the outset that the amount of reinforce-ment needed in an *in situ* concrete pavement is never anything like the amount of steel which might be seen during the construction of a bridge or the floor slab in a building. Unlike such structures, a road slab has support below the whole of its underside. The tensile strength necessary to spread imposed loads over a sufficient area of the subgrade is not great – indeed with lighter loadings and stronger subgrades, the *in situ* slab will often have sufficient tensile strength without reinforcement.

Steel reinforcement might be included not to increase strength in use but for an entirely different reason. Concrete contracts as it sets. It also expands and contracts with temperature variations. It is most unlikely that the stresses which result from expansion and contraction are evenly distributed within the slab – that could only be the case if the concrete was entirely homogeneous and with no variation in thickness. An irregular distribution of stresses can lead to the slab cracking, if not immediately, then when in use. This phenomenon has to be controlled by limiting the size of slab laid at any one time (see below) and can be eliminated by including light reinforcement within the slab to distribute evenly the forces resulting both from contraction and use.

The reinforcement used is covered by BS4483:1985 'Steel fabric for concrete reinforcement'. The term 'steel fabric' might need some explanation. Steel rods, some not much more than wires, can be welded together to form a square or rectangular mesh. This is done by machine in a factory and the fabric delivered to site in rolls or large sheets. BS4483 covers a range of fabrics with variation in mesh size and shape, in the rod sizes used, and in the combination of rod sizes used. Only the lightest of this range have an application in *in situ* concrete pavement construction – the others would give far more strength than would be needed.

Structural theory would always place the reinforcing mesh in the lower part of the slab, for it is there that tensile forces will be at their greatest when the slab bends under load. However, there are practical reasons for placing the reinforcement in *in situ* paving slabs nearer to the top. First, there is no doubt that it is easier, and it might be considered better to have the mesh correctly placed in a theoretically less advantageous position than incorrectly placed elsewhere. Secondly, and more importantly, experiments have shown that should a slab crack those cracks will always run down to the reinforcement but rarely beyond it. With reinforcement near to the top, much more of the slab remains uncracked. Thirdly, the consolidation of the concrete which occurs when it is tamped, reduces in effectiveness with depth. Reinforcement near the bottom of the slab is likely to be in poorer concrete than that near the top. Finally, and again importantly, should frost heave occur (and if it does the forces induced in the slab will be greater than those resulting from light traffic loads) bending would be upwards and so reinforcement near the top would be in the theoretically correct place.

As already stated, steel fabric is available with both square and rectangular meshes, and with both the same and different sized rods forming those meshes. A square fabric using the same size of rod in each direction will distribute loads equally in both directions. A rectangular mesh with larger rods to the long axis than the short, will distribute a much higher proportion of those loads in the direction of the larger rod. When an *in situ* concrete pavement is used as a path or road, a reinforcing fabric should be used that will distribute loads transversely from edge to edge. Where the pavement is a more extensive area then a fabric which distributes loads evenly is preferable.

The British Cement Association (BCA) have given the following advice.

Unreinforced slabs 175 mm (7″) thick will be adequate for low densities of commercial vehicles, whilst 75 mm (3″) would be sufficient for pedestrian-only areas. Reinforced slabs for light commercial traffic could be 125 mm (5″) thick, but reinforcement in pedestrian-only areas would not lead to slab thickness reduction for at 75 mm (3″)

thickness is at its practical minimum. A reinforced road might use a fabric weighing between 1.5 and 2.5 kg/m² (3½ – 5 lb/sq. yd). If a pedestrian-only area were reinforced to eliminate cracking, the lightest available fabric would be satisfactory. Where a high standard of workmanship, sub-base preparation, and concrete is to be expected these figures for the thickness of roads might be reduced by 25 mm (1"). But, 100 mm (4") should be considered as the practical minimum for a reinforced pavement that will take any vehicular traffic; 150 mm (6") for a similar pavement unreinforced; and 75 mm (3") for a pedestrian-only area. It should be noted that an unreinforced slab is not recommended in areas where a base is essential.

By way of examples, the previous table gives descriptions of steel fabrics taken from the trade literature of British Reinforced Concrete Engineering (BRC).

Reference number	Mesh size in mm (in) Major × Minor	Wire size in mm (in) Major × Minor	Weight in kg/m² (lb/sq.yd)
A98	200 × 200 (7.9 × 7.9)	5 × 5 (0.2 × 0.2)	1.54 (2.84)
B196	100 × 200 (3.9 × 7.9)	5 × 7 (0.2 × 0.28)	3.05 (6.88)
C283	100 × 400 (3.9 × 15.7)	6 × 5 (0.24 × 0.2)	2.61 (4.81)
D49	100 × 100 (3.9 × 3.9)	2.5 × 2.5 (0.1 × 0.1)	0.77 (1.42)

1.13.6   The concrete

The sources of all materials should be approved and, with larger projects in particular, samples should be submitted for approval and/ or materials delivered with certificates to the effect that they are as specified. The best and most convenient form of specification is to make reference to the appropriate British Standard.

(i) Cement. Ordinary Portland cement is all that is generally required, but in very special cases rapid hardening cement might be useful. Both are covered by BS12:1978. Portland blast furnace cement (used in Scotland) is covered by BS146:Part 2:1973.

Cement should be gauged by weight rather than volume for the amount of air that might be held within the loose powder can vary considerably. One cubic metre of cement weighs 1442 kg (90 lb/cu. ft).

Cements with added pigment are available. If these are used then cement content should be increased by 15% to allow for the weight of the pigment.

(ii) Aggregates. Aggregates are covered by BS882:1983. Large aggregate, say 38 mm (1½″) maximum size, requires better workmanship than does smaller, say 19 mm (¾″) maximum size. This is because the larger aggregate is more difficult, and takes more time, to tamp to a level surface. That having been said, a 38 mm (1½″) all-in aggregate (that is graded aggregate with the largest pieces being no more than 38 mm, and delivered as such without the various sizes of aggregate having to be mixed on site) is typical of that specified for lightly trafficked roads and footpaths. Further, the larger 38 mm (1½″) aggregate might be specified because of its appearance – a large exposed aggregate can be most attractive.

No single piece of aggregate should be larger than 50% of the finished thickness of the concrete. When a slab is to be laid in one layer, that is of little consequence. However, it might be that the slab is to be constructed in two courses (the top, perhaps, containing relatively expensive aggregates that are to be exposed as a finish). Overall slab thickness would have to be 150 mm (6″) or more if 38 mm (1½″) aggregate is to be used in the upper of those courses.

Finally, smaller aggregate is easier to work around reinforcing rods and so gives a better product. However, with the mesh sizes that are likely with work of the kind considered here, that would not be a major consideration.

(iii) Water. It is best to use water from a drinking source. When specified in those terms it can be certain that the water will be clean and free of any dissolved pollutants. If a drinking water supply is not available, the water should be tested to ensure that it is satisfactory.

(iv) Proportions. Concrete which is to be mixed on site is specified in terms of the ratio of cement:fine aggregate (sand):large aggregate. It is most satisfactory for this to be done by weight, but it is frequently much more convenient to do it by volume.

When mixed by volume, two specifications are common:

(a) 1 part of cement to 2 parts of sand to 4 parts of large aggregate.
(b) 1 part of cement to 3 parts of sand to 6 parts of large aggregate.

The second of these would only be recommended for very utilitarian work, and not even then if heavy loads might sometimes be expected. The more sophisticated technique of proportioning by weight allows more sophisticated specifications.

(Note: when concrete is not to be mixed on site but ordered from a ready-mix company, it is best to leave responsibility for the detail design of the mix with that company. So, concrete in these circumstances should be specified by large aggregate size and strength. The latter is usually given as 21 N/mm^2 (3000 lb/sq. in) after 28 days, for work of the kind discussed here. In addition, any specifics necessary to allow intended finish must be included in such a specification.)

The amount of water that is added to the mix is also critical. If there

		Mix proportions by weight – cement:sand:aggregate			
		Crushed rock		Gravel	
Method of compaction	Maximum slump	38 mm(1½")	19 mm(¾")	38 mm(1½")	19 mm(¾")
By hand	50 mm(2")	1:1¾:3½	1:1½:3	1:2:4	1:1¾:3½
Lt machine	25 mm(1")	1:1¾:4	1:1½:3½	1:2:4½	1:1¾:3¾
Hvy machine	13 mm(½")	1:2:5	1:1¾:4	1:2¼:5	1:2:4½

is insufficient the necessary chemical reaction will not properly occur. If there is too much there would be the danger that the various sizes of aggregate would separate within the mix when it is tamped (rather like a poorly made fruit cake). Contractors tend to like more, rather than less, water in a mix for the concrete is then much easier to handle. It is to limit that tendency that maximum slump is specified in the table above. Once the right amount of water for a particular mix has been established (through the test described below) it is expressed as the volume to be used with each kilo (or pound) of cement; this proportion is called the water:cement ratio.

Simple but special equipment is needed to measure slump. The process is not unlike making a sand castle with a child's bucket and spade, but in this case the 'bucket' is of a precisely specified size and form and the 'spade' is a heavy rod. An initial judgement has to be made about the amount of water that will be needed, and a sample mixed. A 'sand castle' is then made with that sample – having ensured that there are no air pockets in the sample by tamping it with the rod whilst still in the 'bucket'. If the sample, when turned out on to a clean level surface, slumps more than the figures given above, then it contains too much water.

It must be remembered that heaps of both the large and small aggregates are likely to contain water. This can affect the amount of water that should be added on one day compared with the last, and with one batch of aggregate when compared to the last. The best results will be given when slump tests are conducted at regular intervals.

(v) Additives. Additives may be included with the water. They are likely to fall into one of four main kinds. Air-entrainers encourage the formation of minute air bubbles in the concrete and make it much easier to place, particularly if water content is low. In theory there should be reduction of strength but improved workability can offset that loss.

Accelerators increase the rate at which the concrete hardens and can be useful when air temperatures are getting low (but not freezing). Retarders are similarly useful in hot weather. Waterreducers help to disperse cements more effectively in water and so improve workability with low-slump mixes.

All additives are likely to change the surface colour of the slab and so, if they are to be used at all they should be used consistently.

(vi) *Formwork.* Concrete, when it is placed, is a thick liquid. It must be contained precisely in the place where it is required, until it has set. Fresh concrete is also heavy and, moreover, it has to be tamped with heavy equipment. The formwork which temporarily contains the setting concrete, must be exactly where it is wanted, and be strong enough for its job, or there will be disaster.

The formwork most commonly employed with *in situ* concrete pavings is made of steel with a width that is the same as finished slab thickness, has props behind it to ensure it does not move outwards as concrete is placed, and has clips accurately to join one length to the next. Such a system will allow precisely-cast concrete slabs, and has the additional benefit that the forms can be cleaned and reused for the next section. Their disadvantage is that they can get damaged as they are moved about the site or between sites. One survey showed that one-third of the faults reported in concrete roads were due to damaged or faulty formwork.

Formwork can be constructed from timber or plywood, securely located and propped. Adequate strength is more difficult to achieve with timber, and timber forms could only be reused a few times, if at all. However, being made to measure, the designer is not limited to the thicknesses and curves covered by a steel system.

Thirdly, some type of permanent formwork might be considered. Kerbs and edgings are discussed in a later section and some of these would allow their construction prior to laying the slab. But it must be remembered that they might need more resistance to overturning when containing the wet concrete than they will when in use; that they might get soiled during concrete construction; and tamping the concrete would be awkward if not impossible if the upper level of the edging were not also the upper level of the slab.

(vii) *Compaction plant.* The *in situ* concrete slab has to be tamped immediately after it is placed and before the concrete starts to set. This is necessary for much the same reason as the base of a flexible road is consolidated – it gives the maximum mechanical bond between the variously sized aggregates. Further, with a concrete slab, as much as possible of the air that will inevitably be within it, must be removed if the slab is not to be weakened.

Three different pieces of equipment might be employed. A hand tamper consists of a heavy baulk of timber, longer overall than the width of the slab to be finished, steel faced on one of its narrower sides, and with a handle at each end on that side which is not steel faced. The concrete is placed to a level slightly higher than the top of the formwork. Two people take one end of the tamper each and bounce it up and down on top of the wet concrete bringing its surface down to the level of the top of the forms. Each bounce is made fractionally further across the untamped surface. A light machine has a vibrating unit fixed to the top of the hand tamper. This machine only has to be 'walked' across the surface and not lifted and dropped. A heavy machine uses a different approach whereby vibrations are induced in a heavy steel plate.

Whilst compaction is necessary, too much will seriously weaken the concrete. Water will be forced to the surface where it can physically damage the slab as it runs off. More importantly, it will bring cements with it leaving too little within the slab and giving a surface crust which will rapidly deteriorate in use or weather.

(viii) Frosty weather. Water which has frozen is not accessible to the chemical reaction. Further, that reaction is significantly slowed when air temperatures are down towards freezing. So concreting should stop when a falling thermometer reaches 4°C (38°F) and not start again until the rising temperature passes 2°C (34°F).

(ix) Curing. Water must not be allowed to evaporate in any quantity or it will not be available in the reaction. Covering the tamped surface with polythene sheets is the commonest practice, with these being on a lightweight frame when a specific surface finish is to be protected. A more traditional approach is to cover the slab with damp hessian which might have to be sprayed with water should it dry too quickly. Alternatively, the surface of the newly placed concrete might be sprayed with a proprietary curing agent which inhibits evaporation, and this is probably the most useful with extensive areas.

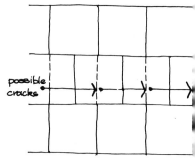

WRONG

1.13.7 Expansion and contraction joints

It has been explained that concrete contracts as it hardens. And, in common with other materials, it expands and contracts with temperature variations. These movements set up stresses within the concrete and the slab will crack if these build up beyond the point which the concrete can resist. Cracks in concrete pavements not only look unattractive but are also likely to lead to premature deterioration. *In situ* concrete pavements must be constructed in such a way that movements can be accommodated – the stresses relieved before they can cause strain. Appropriately designed joints must be included in the slab. The paragraphs which follow consider the design of the joints themselves. But it must also be appreciated that the bays within the slab as a whole must be configured in such a way that movement in one does not cause cracking in those adjacent. Figure 1.7 illustrates this point.

The contraction which occurs as the concrete hardens is the easier to deal with. To avoid confusion with the joints needed to accommodate

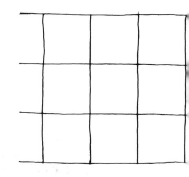

RIGHT.

1.7 The bays within an in situ *concrete slab must be constructed so that movements in one will not induce stresses (and thus cracks) in those adjacent.*

50

movements with temperature changes, these joints are generally known as construction joints. There are three basic methods.

(i) Alternate bay construction. A very high proportion of initial contraction happens in the first hours of a concrete slab's life. So, a slab can be constructed not as a continuous sheet but in a series of bays; alternate bays being laid at first, and the remaining bays cast after the first have hardened. The concrete in adjacent bays abuts but remains separate. This method may slow down the work and could be awkward if it led to much cutting of reinforcement sheets to fit bay size, but it can be quick and economical particularly with unreinforced linear slabs such as footpaths and drives, when bay length might be in the order of 4.5–5.0 m (15′0″–16′6″) (Figure 1.8).

(ii) Dummy construction joints. These allow the slab to crack, but on predetermined lines. A weakness is deliberately introduced by pushing a 6 mm (¼″) 'T' section into the surface of the slab after it has been tamped. The stem of the 'T' penetrates about one-third of the depth of the slab. The 'T' is removed after the initial set but whilst the slab still contains enough water to lubricate its removal (when the concrete is still 'green'). The slab will crack (if it is going to) on the line of this weakness. The narrow space left by the removal of the 'T' is filled to within 6 mm (¼″) of the surface with mastic, restoring the waterproof qualities of the slab. The remaining 6 mm (¼″) can be filled with sand if the joint is to be disguised, or with sand and soil if mosses are to be encouraged. Unreinforced bay length should, again, be about 4.5 m (15′0″) but might be at more than twice that interval with a reinforced road or footpath (Figure 1.9).

1.8 *Concrete contracts on setting. Bay size should not exceed 20–25 m² (200–250 ft²) or that contraction may induce cracks. Bays should be laid alternately, allowing at least 24 hours for initial contraction before the second series of bays is laid.*

1.9 *As an alternative to alternate bay construction (Fig. 1.8) a mild steel T-section can be pressed into the surface of the slab as it is cast. The cracks which form after initial contraction will then be in predetermined positions.*

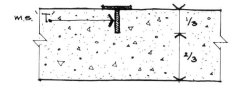

(iii) Sawn joints. This method is quick and neat, but the machines required are expensive and are noisy and dusty in use. The slab is placed and tamped as a continuous sheet and construction joints cut into it after it has hardened but before use. The depth of cut should be about one-third of the thickness of the slab and filled with mastic (see (ii) above). Timing the operation is critical if stresses are to be relieved before they cause strain. This method is really only an option when the work is being undertaken by an experienced and fully equipped contractor (Figure 1.10).

51

Joints to accommodate movement with temperature changes, are generally called expansion joints, although they also have to work during subsequent contraction. Basically, an expansion joint consists of a gap formed in, or between two bays of, the slab. Unlike a construction joint it has to have appreciable width if it is to accommodate likely movement. Such a discontinuity could be a practical embarrassment; for rainwater would erode the base (or formation) below the slab on that line, and differential settlement between the slabs would be much more likely if they were not even connected through friction. So the joints have to be filled with material(s) which can be compressed and will expand again, whilst remaining waterproof, and a way may have to be found to carry reinforcement through the joint without inhibiting its capacity to move.

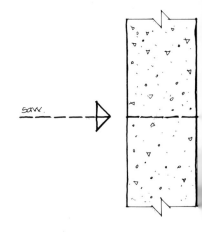

1.10 *A third possible approach to coping with initial contraction in an* in situ *concrete slab, is to saw through the slab after the concrete has obtained its initial set but before contraction cracks have developed where they are not wanted.*

The simplest detail (which would be entirely satisfactory with a footpath) would be to place a piece of softwood, with a thickness of 13–19 mm (½" – ¾") and a width equal to the depth of the slab less 6 mm (¼"), against the edge of every third bay cast, prior to casting the next. (Clearly, this is much easier when the alternate bay method of forming construction joints is being used.) The timber needs to be literally soft, and not just of a softwood species. It should be suitably pressure impregnated.

This detail can be refined by giving the softwood a width that is 19 to 25 mm (¾" – 1") less than the depth of the slab. Mastic and sand can then be used to fill over the softwood strip, as described for 'T' bar construction joints (Figure 1.11). Where heavier traffic loads are anticipated, and particularly if the slab is reinforced, yet more sophistication is needed. Bitumen impregnated fibreboard might be used instead of softwood; or, one or other of the proprietary synthetic rubber strips or compounds made specifically for this purpose.

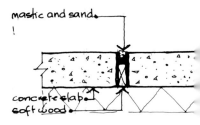

1.11 *In situ* concrete slabs *must be allowed to expand and contract with temperature change. This detail (suitable for a footpath) illustrates that the expansion joint must remain waterproof if the base is not to be eroded.*

Should it be necessary to avoid all risk of differential settlement, steel dowel bars will have to be incorporated across the expansion joint. Half the bar is cast into the edge of one slab before the adjacent bay is cast. The protruding half of the bar is given a coat of bitumastic paint (or similar) so that concrete will not bond with it. On the end of the projecting bar is placed a plastic or cardboard cap the closed end of which being partially filled with compressible material. All this allows the slabs to expand and contract horizontally but stops any differential vertical movement.

Dowel bars in slabs 150–175 mm (6"–7") thick would need to be 19 mm (¾") diameter, 500 mm (20") long, and at 300 mm (12") centres. Dowel bars in slabs 200–225 mm (8"–9") thick would need to be 25 mm (1") diameter, 600 mm (24") long, and at 300 mm (12") centres. Again, the proprietary filler should stop at a point which is the width of the joint plus 6 mm (¼") below the surface of the pavement, so allowing space for a waterproof mastic (Figure 1.12).

It should be noted that any inspection chamber, manhole or other service access covers, or rainwater gully grating, will itself be a weakness in the concrete slab, and so encourage cracking. This is

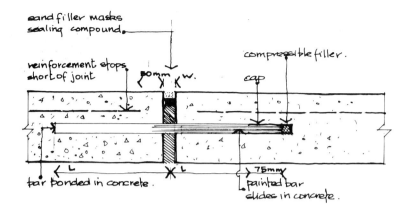

sand filler masks sealing compound.

reinforcement stops short of joint

80mm w.

compressible filler.

cap

bar bonded in concrete.

L

X L

75mm

painted bar slides in concrete.

1.12 *The detail shown in Fig. 1.11 would not be satisfactory in a reinforced* in situ *concrete road. Loads must be shared between bays if the surface of the road as a whole is to remain consistent. This detail shows how this can be done.*

emphasized by the fact that the slab will be partially supported by the walls to the chamber whilst the remainder is supported by the base/subgrade. It is unlikely that both the chamber and the subgrade are providing identical degrees of support and so traffic loads are very likely to open up any potential cracks resulting from expansion and contraction. It is best to run at least a construction joint around any covers in an *in situ* concrete road, at a distance from the cover which places them beyond the walls to the chamber below (Figure 1.13).

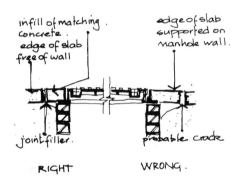

infill of matching concrete.
edge of slab free of wall

edge of slab supported on manhole wall.

joint filler.

probable crack

RIGHT

WRONG.

1.13 *An unreinforced* in situ *concrete slab must not be allowed to have inconsistent degrees of support, or it will crack. This can be a particular problem at inspection chamber covers and the like. This detail shows how that problem can be avoided.*

One final point on construction and expansion joints. Rarely would it be the case that practicalities alone dictate where such joints should be run. The designer will always have some opportunity to judge the matter aesthetically and should always take that opportunity. Shapes might be formed which, through family relationships, give a consistent character to the whole length or area of pavement; variation in bay size might give modulated pattern; formality or informality might be enhanced through bay geometry; scale or directional emphasis might be adjusted. But care must be taken that aesthetic concerns do not override practical considerations. Expansion joints in particular (construction joints to a lesser degree) must run through an area of *in situ* concrete pavement; they cannot step or stagger without the risk of inducing cracks in neighbouring slabs, adjacent to the places where they change direction.

1.13.8 Finishes

The many different concrete finishes available fall into one of two classes: untextured and textured. The latter can itself be divided into two parts, textures given mechanically to the surface, and textures resulting from exposing the aggregates.

(i) *Untextured finishes.* There is, in fact, no such thing as an untextured surface. If nothing else is done, the concrete will always show the tamping marks. These, in an informal context and provided that they are reasonably parallel and of even width, can look well. In such circumstances consistency will be all important. It would be well worth while having the contractor make a sample panel which, after approval, becomes the model for the whole. Some colour variation is possible, given in the main by the colour of the sand used in the concrete mix. Initially the slab will have the bright pale grey of cement but as surface laitance is worn away, and as the slab weathers, the brightness disappears and something of the colour of the sand will augment the grey. Pigments might be added with the cement but these should be used carefully if the effects are not to be crude. All will reduce in chroma over time.

(ii) *Mechanically applied textures.* The application of mechanically textured surfaces to *in situ* concrete pavement is primarily done for utilitarian reasons, the most common being to give a grip, for feet and/or tyres. When done solely for such reasons, the results can be aesthetically haphazard. However, given some thought and care, both practical and visual considerations can be satisfied.

A light-weight roller can be taken across the surface of the concrete when it is still green. These rollers can have one of a number of surface textures and this is transferred to the concrete. The commonest texture is given by a roller covered with small cones or pyramids giving an all-over indented surface. Each indentation is about 10 mm in size and with a similar width of the original surface remaining between each. Other rollers give a texture made up of interlocked, flat lozenges, and others a cross pattern, slightly resembling small setts. There is almost no limit to the textures which could be applied in this way (the practical limit is that the roller must not pick up the surface of the green concrete). However, an extensive area of concrete would be needed to justify the cost of making a roller that would give an individual texture for a particular project.

Textures can be given to the surface by a method not dissimilar to the way in which the pavement was tamped when first placed. The board is much lighter and has a square, chamfered, 'V' shaped or rounded edge. Again two people are best (one at each end) to ensure the grooves which are made are parallel and evenly spaced. This method can be particularly useful on ramps, with the grooves running diagonal to the ramp, or in a 'V' arrangement running from the centre out, and thus draining rainwater to the sides. Individuality can be given to a specific project fairly cheaply in this case. The cross-sectional profile of the board could be easily modified, as could be depth and interval of texture. Adjacent bays might be textured in contrasting directions. The board might also be given a longitudinal profile. But, as before, a sample should be approved before proper work starts, and the designer might want to experiment at that preliminary stage.

Mechanical textures can also be given to the surface of green concrete by laying on a patterned steel grid and applying weight. There are firms who will do this on a subcontract basis with some of the grids they use giving textures reminiscent of stone flag pavings, cobbles, fan pattern sets, or whatever. Somewhat paradoxically, those which are less similar to the originals seem to be more effective – perhaps a good imitation of texture draws attention to the fact that other qualities are wrong. What is certain is that there is no need to be ashamed of *in situ* concrete pavings – they have enough potential not to have to pretend to be something else.

(iii) Exposed aggregate finishes. There are a number of ways in which the concrete's aggregate can be exposed to give both colour and texture. These are discussed later. First a number of general points should be made. (Many of these notes apply also to precast concrete and the surfaces of *in situ* concrete walls, bridge abutments, etc.)

The colour of an exposed aggregate finish depends not only on the colour in the aggregate itself but also on the finished texture of the surface. That, in turn, depends upon aggregate size and variation of size, and on the degree of exposure. It is therefore important to inspect and approve a sample of the proposed finish; simply selecting the aggregate will not be enough. The finer the texture the lighter will be the colour of the surface tone. This is because the texture will produce small areas of shadow and shade within it, and the more extensive these are the darker will be the resulting tone. The British Cement Association (BCA) have an excellent selection of sample panels at their headquarters showing some of the hundreds of different colours and textures available, and the effects of time and weather upon them. BCA will also give technical advice on methods of exposure and, where necessary, send experts to sites to demonstrate techniques.

Colour will depend on precise specification: the shape, size, colour, grading, and degree of exposure of the aggregate, and, to a varying extent, the colour of the cement.

The effect that any difference in the colour of the cement will have, is also dependent upon the aggregate. If a well graded aggregate is used the colour of the cement will affect only general tone whilst the aggregate remains dominant. With large, single sized aggregate with a shallow depth of exposure, the colour of the cement will be perceived as a background to the larger texture, particularly if the cement and aggregate colours are selected to contrast. White cement should always be used with light coloured fine textured aggregates where an overall lightness is the objective.

With some techniques for exposing aggregates, a surface laitance of cement will, at first, colour the slab. This can be washed off with a 10% solution of hydrochloric acid, if there is not time to wait for use or weather to do the job.

Cements coloured with pigments should be used with great care, or harsh, unsubtle colours will result. The cement should not be thought of as a colour in its own right, but in combination with that of the aggregates.

The range of colours available in aggregates are many and various – and there can be a variety within one quarry. The list below is, then, indicative only – as always, samples must be inspected and approved.

Granites	ABERDEEN:	pink and grey
	LEICESTERSHIRE:	pink
	MOUNTSORREL:	reddish brown
	HINGSTON DOWN:	light grey with a touch of brown
	CRIGGION:	greyish green
	PENLEE:	dark blue-grey
	PALESTINE:	light grey with black flecks
	SHAP:	blue almost black
Limestones	DERBYSHIRE SPAR:	white
	DERBYSHIRE CALCITE:	white
	PORTLAND:	soft white
	STONEYCOMBE:	pink
Marble	Great variety,	highly decorative
Slate	WESTMORLAND:	green
	WELSH:	blue-grey and purple-grey
Flint		brown, grey and black
Basalt		blue and black
Gravels/shingles		great variety

Granites, flints and gravels can show variety within the individual pieces. Granites and spars can show sparkle. Flints can have a sheen. Gravels and shingles, when formed from certain metamorphic rocks such as marbles can prove highly decorative, but their ovoid form does not give strong concrete.

Cost, as always with aggregates, will be significantly affected by transport charges.

Types of aggregate suitable for road construction are covered by BS882:Part 2:1973. The need for resistance to wear must be considered and the BS rejects softer materials such as friable sandstones, chalk, soft limestones, shales, soft schists, and broken brick. The need to provide skid resistance must also be remembered. However, the BS is dealing with aggregates for highways and in areas where only light traffic (if any) is to be anticipated, some of these might be satisfactory. In pedestrian-only areas, for example, an exposed broken brick aggregate can be used with great success. Alternatively, problems of both cost and the strength of the main slab might be overcome through two-course construction, with an aggregate selected for the main slab on purely practical grounds, whilst that in the top course is chosen for its qualities after exposure. (It is important that the top course is laid immediately after the main slab, to ensure a monolithic structure.) But even then, very small rounded gravel or shingle should be avoided for pavements – it gives little or no grip for tyre or shoe and its form means that it tends to kick loose.

The concrete mix used when aggregate is to be exposed will vary depending upon desired results, but it would be best to stay within the range of mixes outlined in section 1.13.6. A gap-graded mix can be very effective, and in particular with two-course work where any reduction in strength is unlikely to be critical. A graded mix has been described as an aggregate with a range of sizes from maximum to minimum designed so that each size will consolidate into the interstices between the larger pieces. A gap-graded mix omits one or two sizes from the range. For example, a normal mix might have large aggregate ranging down from 19 mm (¾") through to 3 mm (⅛") and fine aggregate ranging down from 3 mm (⅛") to dust; whilst a comparable gap-graded mix might have the same fine aggregate but have no large aggregate smaller than 6 mm (¼"). With such a mix, a larger aggregate will have more dominance, after exposure.

(iv) Methods of exposing aggregates. Whilst the techniques for exposing aggregates are not, in themselves, complicated they do need a deft and experienced hand if full potential is to be realized. The inexperienced tend either to be too timid or too bold. In the first case, little aggregate but a lot of cement is exposed. In the second the aggregate is left without adequate connections back to the slab and soon kicks loose. Two broad categories of techniques are available:

(a) A slab of appropriate mix is laid as described earlier. A few hours later (a period that will vary with temperature) when the concrete starts to stiffen, a stiff-bristled brush is used to take away a proportion of the cement and fines at the surface, until desired texture is achieved.

A minimum of water has to be hosed on to the slab as this is done to wash away the material being brushed off; but it cannot be stressed too highly that only a minimum of water should be used. Later, when initial set has been achieved (say the next day) it should be brushed and hosed again. As the concrete is now much harder, more vigour and more water can be used. When well hardened (say a week later) a 10% solution of hydrochloric acid might be used to remove any laitance. Whether or not that is done, the pavement should not be brought into use for at least 28 days, and it should be well hosed and brushed immediately beforehand.

(b) Again a slab of appropriate mix is laid, but in this case left until well hardened (at least a week and longer would be better). One of a number of different machines is then used to abrade the surface, taking away the finer softer material. The principal advantage of this approach over that described above, is that the amount of exposure required can be precisely judged and consistently achieved. The principal disadvantage is that it is more expensive. The two commonest ways in which this is done are grit-blasting (when abrasive is projected at the surface by a high-speed blower) and wire-brushing with a circular brush flexibly connected to an electric motor.

There are several other ways in which a texture might be applied/given to an *in situ* concrete paving, but all have certain deficiencies which might make them inappropriate other than in very special cases. These will, then, be only broadly described.

A very smooth surface can be achieved by floating the slab immediately after placing and tamping. A rectangular tool of either wood or metal is used to work up the surface and remove irregularities. A wood float will give a less smooth surface than will a metal one. In both cases an unfloated margin can be left to each bay. However, the degree of smoothness achieved would reduce, if not remove, any non-slip characteristics.

An exposed aggregate finish can be achieved by spreading the aggregates which are to be seen over a newly laid slab, and tamping these into the surface. The height of the formwork has to be adjusted before that finish is tamped. Whilst this technique can give a very bold texture the aggregate can relatively easily be picked off and the larger it is the more likely that is to happen.

Mechanical hammers with a variety of sized and profiled bits, can be used to tool the surface. Most of the effects that can be achieved on stonework can be given to concrete in this way. It must be noted that it is not the aggregate in the concrete that is tooled (unless it is very soft) but the surface of the slab itself. Deep textures, then, involve hammering out some of the large aggregate. This technique is labour intensive, making it expensive.

A final point needs to be made in this section. It is most likely that bay sizes would have to be less than those required for technical reasons, if any surface finish is to be given to wet or green concrete, other than by tamping; for it must be physically possible to reach the whole of the surface to each bay. However, this need affect only the distribution of construction joints and not expansion joints.

1.14 INTERLOCKING BLOCK PAVEMENTS

Interlocking block pavements must be classified as a form of flexible construction for these structures, when considered as a whole, have no theoretical tensile strength. But, the blocks which are used as the surface course do, in themselves, have tensile strength which means that their design can be approached in a way which shows significant differences to the design of macadam pavements. Moreover, the kinds of labour and materials used in their construction, and the visual qualities of the resulting surface, are radically different from other flexible structures. All that being so, it is both convenient and logical to consider interlocking block pavements as one of the three major types of road construction; bound macadam and *in situ* concrete being the other two.

1.14.1 Uses

Interlocking block pavements could be used in any circumstances in which macadam or *in situ* concrete might also be considered, other than for roads where traffic speeds in excess of 50 km/hr (30 mph) are anticipated. They might be found preferable to bound macadam when it is desirable to limit overall depths of construction and avoid the use of specialized and/or heavy equipment. They might be found preferable to both bound macadam and *in situ* concrete structures when a surface pattern and texture is wanted which reduces the scale of the surface as a whole. They might be considered less satisfactory than the alternatives for small areas of very lightly trafficked drives and pedestrian-only surfaces (particularly if these were very irregular), but even then they might well be used to give consistency where those small areas were but a part of a larger scheme using interlocking blocks elsewhere.

1.14.2 The materials

Interlocking blocks are available in both concrete and clay. When viewed from a technical position, there is but one difference between these two materials; thickness for thickness, concrete is stronger. So, a 90 mm thick clay block might be needed in cases where a 65 mm concrete block would suffice. Clay blocks tend to be more expensive than concrete but some might argue that the actually and psychologically perceived qualities of clay gives them more aesthetic potential.

Most clay blocks and many concrete blocks are rectangular on plan with much the same proportions as a brick. Rectangular concrete paving blocks are 200 × 100 mm in format with thickness varying between 60 and 100 mm. They may have chamfered upper edges.

Rectangular clay paving blocks come in a wider range of sizes and thicknesses. The metric equivalent of the traditional building brick is available but because its size, 215 × 103 × 65 mm, has developed over time to give dimensional co-ordination when building walls with mortar joints, it is less convenient as a format for butt-jointed pavings. 210 × 105 × 65 remains close to traditional size but establishes the 2:1 dimensional relationship that is more useful with butt-jointed flexible pavements. 290 × 90 × 90 & 65, and 190 × 90 × 90 & 65 as well as 200 × 100 × 90 & 65, are also available. A block with a long dimension that is not an exact multiple of the length of its shorter side can only be used in running bond which gives less strength than some other patterns (see below). It should be noted that there is an even wider range of sizes and thicknesses available in brick paviers. These can be laid with a very similar technique to that which will be discussed here. But most are relatively thin and many have low strength making them suitable only for footpaths and similar areas, when used flexibly, or as a surface course to rigid pavement (see section 1.13.8). Again, clay interlocking blocks may have chamfered upper edges.

Most manufacturers of interlocking concrete blocks also make non-rectangular shapes. These fall into two classes. There are those which are made as irregular modules in order to increase side length and so increase bond strength in the pavement as a whole. There are others where a number of shapes and sizes are included within a system in order to allow a range of different patterns to be created. These latter systems also tend to have less thickness and so are more appropriate for lightly trafficked and pedestrian surfaces. Both functional and decorative concrete block systems (if we could be allowed to make that rather simplistic distinction) include special shapes to minimize cutting at edges, bends, junctions, etc.

Some clay block manufacturers also make non-rectangular shapes. But differences in manufacturing processes make clay more limited in this respect. Again specials are available for paving edges, etc.

1.14.3 Colour and surface texture

Whilst concrete interlocking blocks are available in a wider range of hues than are their clay equivalents, the clay blocks have more tone and more variation of colour within a single block. Clay blocks are generally in the red/brown sector of the spectrum, some with over-tones of blue/purple and some at the yellow/buff end of that range. Concrete interlocking blocks (if not self-coloured) have added pigments which may fade with time. These include charcoal and paler blue/greys, browns and red/browns through to buffs, and some high chroma colours such as 'marigold'.

Most concrete interlocking blocks have a relatively smooth surface texture but a few are more rugged in appearance. Clay blocks generally have more texture than concrete but none are as rugged as the few mentioned above.

Designers can consider mixing colours within one area of paving. This can be done to some effect when there is a practical objective behind the idea. Areas for parking rather than for through traffic, divisions between parking spaces, running bond drainage channels within a herringboned space, are all examples where colour change might be prompted by functional criteria. But changes made simply for changes' sake can look pretentious.

Whilst fully accepting that tastes vary, the authors feel bound to say that to their minds concrete blocks with high-chroma pigments should be avoided in all but the most rumbustious of circumstances. They would make a particular plea that designers respond to the qualities that concrete undoubtedly has, without requiring it to pretend to be something else.

1.14.4 Edgings

Kerbs and other edgings are both functionally and visually important with all forms of pavement but, as will be seen, with interlocking blocks they are critical. Both concrete and clay paving block manufacturers make special units which will satisfy this critical function, as well as the general criteria applicable to all kerbs and edgings. Kerbs and edgings generally are covered in section 1.17.

1.14.5 Technical design

All interlocking block pavements will consist of three layers, and the more heavily loaded will have four. Subgrade and formation must be designed as described for bound macadam pavements in section 1.6.5. On that formation will be a sub-base, laying course and surface course. With roads and other primarily vehicular surfaces that will take heavy loads and are intended to have a very long life despite those heavy loads, a base will also be included. Determining whether or not an interlocking block pavement needs a base is relatively easy.

Experimentation has shown that interlocking blocks, when correctly constructed on a sand laying course, will have a load-spreading capacity equivalent to the surfacing and base of a bound macadam road of the following overall thickness: 225 mm when the base is lean concrete; 170 mm when the base is dense bound macadam; 160 mm when the base is hot-rolled asphalt. So, if the calculation as described in section 1.7.2, using Figure 1.5, shows that the depth of construction needed for a bound macadam road (excluding any necessary sub-base depth) is no greater than these dimensions, then an interlocking block pavement could be constructed without a base. If a base is needed it should have a thickness given by subtracting the thickness of the blocks and their sand bed from the necessary construction depth of the equivalent bound macadam (again excluding sub-base depth), subject to a practical minimum of 100 mm. Sub-base thickness will be the same irrespective of whether it is a bound macadam or interlocking block pavement that is to be built above it.

A change of texture shows the way with a subtlety which enhances contrasts between paving and buildings.

By way of example let us assume that trial holes have disclosed a clay subsoil (CBR 3%) and the amount of traffic anticipated suggests the use of curve B, then an overall depth of construction of 375 mm (15″) is shown to be needed. If it is assumed that available sub-base material will have a CBR of 10% when consolidated, then the graph shows that 225 mm (9″) of bound macadam would be needed above a 150 mm (6″) sub-base. Such a pavement could be designed as a bound macadam road with a 65 mm tarmac surfacing over a 160 mm lean concrete base, and 150 mm consolidated sub-base. But interlocking block pavements have a strength equivalent to a tarmac surfacing over a lean concrete base with a combined thickness of 225 mm (9″). So, in this circumstance 80 mm (3.5″) concrete interlocking blocks over a 50 mm (2″) consolidated sand bed, over a 150 mm (6″) consolidated sub-base, would be one alternative specification. However, had only slightly higher traffic densities and/or slightly lower bearing capacities been assumed, then the interlocking blocks would have had to be given a minimum 100 mm (4″) consolidated base over that sub-base. (Quite high traffic densities and low bearing capacities were deliberately selected for this example in order to show the probability that interlocking block pavings would not require a base when used for the kinds of work this book is intended to cover.)

Care must be taken, however, to ensure that the thickness of construction provided is adequate to cater for the requirements of frost-susceptible sub-grades and sites with higher water tables. In these respects the design of interlocking block pavements is as for other flexible structures (see section 1.6.5).

As with almost every other type of paving, interlocking blocks should be built to have suitable falls to a surface-water drainage system. Whilst when they are first constructed they do allow a certain amount of percolation through the minute joints between them, those are soon totally sealed by use. Rather more care has to be taken in this respect than is the case with either bound macadam or *in situ* concrete pavements. Complex patterns of falls are more difficult for the individually rigid blocks to accommodate. In order to make its point, Figure 1.14 exaggerates the degree of fall that there might be between two adjacent areas of paving, but it does illustrate how mechanical bond and visual appearance would both be weakened in such circumstances. It is for this reason that with extensive areas of interlocking blocks, it is recommended that drainage should be to channel blocks or drains, rather than to individual gulleys. With linear constructions such as roads and footpaths, the problem is more easily overcome through cambers or cross-falls to channel blocks at the kerb which themselves fall to gullies. Cross-falls of 1:40 (2.5%) and longitudinal falls of 1:200 (0.5%) minimum to 1:100 (1%) are recommended.

Thought should be given to the sizes and locations of inspection chamber and other service access covers, gully gratings, etc. With other forms of pavement, visual quality is all that is likely to be

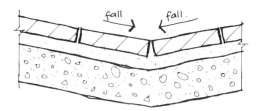

a. WRONG – interlocking block omitting channel block or drainage channel.—

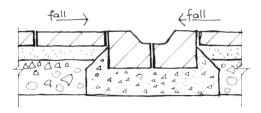

b. interlocking block using propnetory channel block

1.14 *Interlocking concrete block pavements must be given properly designed and constructed channels. Simply allowing two areas to fall towards each other will give inconsistent support on the line where they meet.*
Note: *The degree of fall has been exaggerated in this figure in order to make the point.*

impaired if this is not done. With interlocking blocks absence of such thought would lead to unnecessary, wasteful, and weakening cutting of the basic units. It could even lead to having to fill gaps with *in situ* concrete because their sizes are too small (that is, any gap less than one-quarter of a full block). Moreover, the effect on the visual quality of interlocking blocks resulting from a lack of such consideration could be rather worse than with other forms of pavement, for it would not only be the covers that appeared haphazard but, in all likelihood, the bonding arrangements as well.

There is value in giving a block border to service access covers etc., particularly when using a herringbone or similar bond. Otherwise the blocks immediately adjacent to the cover will receive some support from the foundations to the chamber beneath and slight differential settlement might be experienced. With a border to the covers any differential settlement will be less obvious and the risk of individual blocks being displaced, or cracked, is avoided. But such a border will put more visual emphasis on any poorly located cover, making it yet more important that the covers themselves are considered as an integral part of the design.

1.14.6 Block thickness

The following block thicknesses are recommended for concrete interlocking pavings:

Heavy industrial traffic   80 or 100 mm

Residential roads	60/65 or 80 mm
Car parks and similar	60/65 or 80 mm with some heavier use
Pedestrian-only areas	60/65 mm

It will be remembered that clay blocks are generally less strong than are concrete, and so these thickness figures might have to be increased by 25% in all but pedestrian-only and very lightly trafficked areas. As should be clear, clay blocks would not really be suitable for industrial areas other than by (somewhat uneconomically) using them on edge in order to give them a greater thickness. On the other hand, in pedestrian-only areas the clay paviers discussed in section 1.15 which are thinner than the thinnest concrete block, might show advantage. There seems to be a certain logic in these observations in so far as concrete as a material associates well with the rugged purposefulness of (say) a quay or dockside, whilst the visually softer clay paviers might be found more suited to a casual, verdant context.

1.14.7 Visual considerations

All interlocking blocks must be laid to some kind of bond, but some allow a wider range of bonding patterns than do others. As has been said, blocks which are not dimensionally co-ordinated can only be laid in running bond. Other rectangular blocks can be used in this way if desired.

Rectangular blocks which are dimensionally co-ordinated can be laid in ways which improve mechanical bond (and so the strength of the road) and, in some contexts, provide a more effective pattern. Herringbone bond is the commonest of these when successive blocks

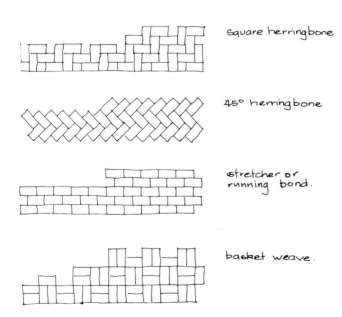

square herringbone

45° herringbone

stretcher or running bond.

basket weave.

1.15 *Interlocking concrete block pavements can be bonded in a number of different ways. This figure does not illustrate them all. Note that the direction of the bond relative to the direction of the traffic will have a considerable effect on the character of the pavement.*

65

are placed at right angles to each other, with the paving as a whole developing as a series of chevrons. With square herringbone, half of the blocks are perpendicular to a kerb whilst with angular herringbone all are at 45° to kerb. Angled herringbone provides a stronger road whilst square herringbone is more static visually and so might be more appropriate in quieter areas. Basket weave bond uses the blocks in pairs with the two short sides of one pair butting against an adjacent long side. Basket weave is less strong than either of the herringbone bonds, and is almost invariably used square to the kerbs where through traffic is not anticipated, or where a sign is to be given to through traffic to encourage care and reduction of speed. Figure 1.15 illustrates these bonds.

Some manufacturers offer a half-block, 100 × 100 mm (4″). This can be of value at the edges of square herringbone, for example. Blocks can easily be cut in half but then it is best to use unchamfered units throughout, for the cut edges will have no chamfer. With a half block

shaped blocks

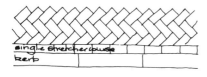

single stretcher course.

double stretcher course ;— note single or double stretcher course against a kerb or abutment not only makes a neat finish but makes the marking of the cut blocks easier.

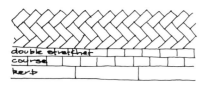

cut / purpose made half blocks ;— cut or purpose made half blocks can eliminate small cuts at abutments when using 45° herringbone.

1.16 *The bonding details of interlocking concrete block pavements at their edges needs to be considered with some care. Most manufacturers include special items within their range to assist with this.*

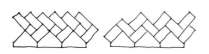

special fittings ;— these are made for use with rectangular blocks laid in 45° herringbone — this eases laying.

many bonding arrangements become possible, and some of these are illustrated in Figure 1.16.

Blocks which have been made non-rectangular for essential functional reasons, have to be laid in the way for which they were designed. Almost all give a version of running bond, but some are closer to basket weave. Only a few, such as the bow-tie shape available from some clay block manufacturers, offer the designer bonding options other than choice of direction in which the bond should run. All of the companies making irregular interlocking blocks offer standard specials to assist with paving at edges, junctions, services covers, etc.

Those blocks which are made irregular for essentially visual reasons offer a wide range of bonding options. This flexibility is achieved by including a range of shapes and/or sizes in the system. That being so, there is less need for specials – given that the designer exploits the potential that is within the system to cope with special circumstances.

1.14.8 Construction

Excavation and preparation of subgrade and formation proceeds as described for bound macadam pavements. Remember that any falls that are to be included in the finished surface must be established in the formation. The sub-base (and base if any) is then laid and consolidated as described in section 1.7.4.

The edgings must then be constructed. With interlocking block pavements the edgings (and sometimes, divisions within an area of paving) are an integral part of the structure. They must be properly designed and soundly built, if the whole of the pavement is not to deteriorate rapidly. Both upstand and flush kerbs are possible, using either blocks of the kind used for paving or some other material. Whatever the material it must be firmly set on an adequate concrete strip foundation (say 200×100 mm ($8 \times 4''$) giving 21N/mm^2 (3000 lb/in^2) compressive strength after 28 days (1:2:4) mix if hand-batched) and thoroughly haunched behind with similar concrete; with that haunching taken as high as the neighbouring surface will allow. Two of the ways in which this might be done are shown in Figure 1.17 whilst kerbs, edgings and drainage channels are considered more thoroughly in section 1.17. It is absolutely critical that the edgings are left for the concrete to harden thoroughly before construction proceeds. If that is not done, vibration of the laying and surface courses will simply push out the edgings, leaving them in the wrong place and the pavement inadequately compacted.

The laying course should consist of sharp sand or crushed rock fines. It should be well graded from 5 down to 0.1 mm, with it being important that there should be not more than 10% of the largest size and at least some but not more than 10% of the smallest. It should contain no more than 3% by weight of clay and/or silt.

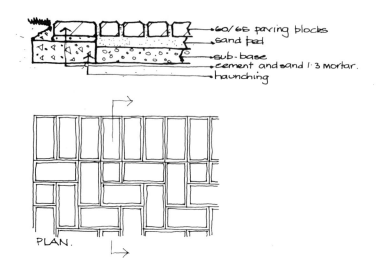

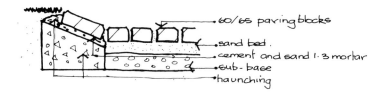

The laying sand should not be very wet and should be supplied and used at as constant a moisture content as proves possible. This last can be difficult to achieve, for any heaps of sand left awaiting use will almost inevitably get wetter or drier depending on weather conditions. Moreover, it is probable that different parts of the heap will develop different moisture contents. A constant moisture content is necessary to facilitate consistent compaction. It is best to cover heaps from rain and wind, to take sand from different parts of the heap when barrowing, and to undertake compaction tests at regular intervals, particularly in changeable weather.

Laying sand is spread over the sub-base or base and screeded level at a thickness that will compact down to 50 mm (2″), give or take a very little. If the sub-base or base has been very carefully laid and shown little surface variation, compacted thickness could be reduced to 30–40 mm (1.2–1.6″). Judging the uncompacted depth that will give desired results can be tricky. Guidance is offered which suggests that 15 to 20 mm (0.6–0.8″) of compaction might be expected when it is a 50 mm (2″) thickness that is ultimately required, and 10–15 mm (0.4–0.6″) when a 35 mm (1.4″) compacted laying course is the objective. But if there is any uncertainty, tests should be conducted.

Screeding, in this case, simply means that the sand is given a totally level surface. This is easily done with a board, longer than the span between edgings, with notches cut from each end that have a depth that is the same as the vertical distance between the top of the unconsolidated sand and the top of the edging. Two people can then scrape the notched board along the edgings and its lower face will take off the sand to the required depth and indicate where more must be added.

Where ultimate strength and/or precision of surface is an important consideration, the laying course can be constructed in two courses. The first to be laid should constitute two-thirds of the ultimate consolidated thickness and should be consolidated to that depth with a plate vibrator before the second course is laid.

It is important that no larger an area of sand be spread than be covered with consolidated blocks during the remainder of that day. Otherwise, there is a risk, at best, of the sand getting very wet should it rain and, at worst, of litter or other undesirable material becoming incorporated in the structure.

The blocks are then laid in the desired bond, directly on to the unconsolidated sand (or upper course if two-course work). This is best and most frequently done by hand, even with extensive areas, although there are machines available that can pick up and place a number of blocks at a time. Each is abutted as close as proves possible to its neighbour. There is no need for the individual blocks to be tapped into place. Labour can work from the blocks already placed, and it is perfectly acceptable to hand-barrow across those unconsolidated blocks in order to supply that labour. Full blocks (those that do not have to be cut) should be laid first, starting at a corner where line and angle can be precisely established. Work should proceed diagonally, whatever the bond, to give a laying face that can be worked from quickly and without having to fiddle individual blocks into place. Regular checks should be undertaken to ensure that lines within the bond that are intended to be straight, are straight. Two people at either end of a lightly tensioned string is the simplest technique.

Other workers should follow up behind the layers, filling the gaps left for cut blocks. (As an aside, it is worth saying that it is useful if the layers and cutters change roles at intervals. Not only will they remain more alert both physically and mentally, but the team as a whole is more likely to develop a pride in the pavement as a whole.) Blocks can be cut with a hammer and bolster, but if there is much cutting to be done a block-splitter should be used. This hydraulically-assisted machine brings a guillotine down precisely where the cut is needed, and so ensures much tighter laying. Pieces that are less than a quarter block are difficult to cut and so designers should avoid that necessity.

No more an extensive area of blocks should be placed at one time than can be compacted in what remains of the day, or half-day. Pavings should not be compacted in small areas, or surface irregularities may develop. An area of blocks should be compacted by making two or three slow passes with a suitable plate vibrator. In this case, suitability is a fairly flexible term – if the machine achieves necessary compaction, it is suitable – a heavier machine might do it with fewer passes but with a small intricate project, a lighter machine might be better. If there are no particular constraints, it would probably be best not to specify the machine to be used, only the result that is to be achieved. Consolidation should not go beyond a line 1 m (3'4") back from any temporary unrestrained edge.

Vibration will bring the blocks down to the required level whilst, at the same time, forcing some of the sand up into the gaps between them. Clean, dry sand to the same specification as for the laying course, but from 1.5 mm (0.05") down, should be spread over the consolidated blocks and swept with a stiff-bristled broom, particularly over and into the joints. One more pass with the vibrating plate and a final soft brush to clean completes the operation. Laid blocks should not be left overnight without top-sanding, nor should traffic be allowed to use the pavement until that is done.

Once the work is completed, it can be put to use immediately. Indeed construction traffic might use completed parts whilst other areas are still under construction – provided that it is kept away from any unconsolidated blocks, laying course, and prepared formations.

There will be some settlement with use, particularly with heavier and denser weights of traffic. This presents no general difficulties for the surface as a whole will settle as one. But it could be embarrassing if the settlement left inspection chamber covers, etc. proud of the surface, or if surface water no longer drained into channel blocks or linear drains. These elements will not settle for they are on their own foundations. The difficulty can be avoided if any elements which are ultimately to be flush with the surface, are initially set down by 2–3 mm (0.08·0.12"). Similarly, any such element that is intended to be lower than the surrounding surface should initially be set down by an additional 2–3 mm (0.08–0.12").

1.14.9   Generally

Interlocking clay or concrete blocks will offer smaller scale and constructional convenience, when compared to other forms of pavement, but their design must be considered to a finer level of detail if those potentials are to be realized.

1.15   SLAB, BRICK, SETT AND COBBLED PAVEMENTS

The earlier sections of this chapter have concentrated their attention on pavements with strengths which allow regular or occasional vehicular traffic. Whilst comments have been made about lighter-weight versions of those structures and their value for pedestrian-only surfaces, those have been essentially as asides. That which follows examines approaches to pavings design which, for a number of different reasons, are likely to be most suited for foot traffic. There will be more asides, but in those cases it will be where there are variations that might allow use by motor traffic when surface consistency is required.

1.15.1 Common matters

There are some matters which are common to all forms, and these are discussed first.

(i) *Foundations.* All the different types of construction covered by this section will, or may have to, have a similar foundation. To avoid repetition these are covered here.

(ii) *Subsoil drainage.* The circumstances which might require subsoil drainage and the ways in which that can be done, are all as described in section 1.6.5. All that needs to be added is that the cost of such drainage could well be a much higher proportion of the total costs of a path, than with a road. If costs were becoming prohibitive it would be better to use one of the alternative forms of construction that are less likely to be adversely affected by poor subsoil drainage.

(iii) *Preparation of formation.* Again, the information given in section 1.6.5 is also applicable here.

(iv) *Base.* Yet again, the information already given in section 1.2.7 should be read in this context. But the loads which the base has to spread are very low. The chances are that the load-bearing capacity of the subgrade will be entirely adequate without the base having to spread the imposed load at all. So, with pedestrian areas the base is provided as much (if not more) as to give a clean convenient surface on which to work, as for its structural strength. 75 mm (3″) of well consolidated materials laid to the falls of the finished surface, should be adequate in all but the most abnormal circumstances. Further, the qualities of those materials is less critical. Broken brick hardcore, for example, or clinker is often used when these might be considered too weak or too frost susceptible for a road. But care should be taken with clinker; it may contain soluble salts that might attack the masonry and/or foundations of adjacent structures.

(v) *Base for occasional motor traffic.* Some of the materials discussed below are, or can be, used as quite small units. In those cases they can be considered as the wearing course to a flexible road construction, given that the foundations are suitably designed (see section 1.7.4). Others have a ratio of breadth to thickness which would make them very likely to crack if heavily loaded whilst not being fully and evenly supported from below. Such materials are not really appropriate for vehicular traffic. But there can be times when vehicles might be expected to cross a small length of footpath, and a consistent surface is required. The most satisfactory decision then would be to build an *in situ* concrete pavement as described in section 1.13, with its surface set at a level that allows the general treatment to be carried over it. The ways in which that might be done are covered in the various parts which follow. (To this should be added the observation that it is not worth while to design pedestrian-only pavements to take the loads of very rare motor traffic. It would, for example, be cheaper to repair a footpath after it had been crossed by, say, a fire engine, rather than to build it initially in anticipation of that possibility.) A 100 mm (4″) *in situ* slab, probably without reinforcement, laid on a 75 mm (3″) well consolidated base, is likely to prove satisfactory in most cases.

1.15.2   Stone paving flags

(i)   Uses. Stone flags are now very expensive to buy, even as salvage, but they remain one of the most durable and pleasant paving materials available. Like other natural materials, they have the advantage of toughness combined with slight irregularities that gives them an individuality that is absent in machine-made products. They can be used in both formal and informal contexts, and either with or without regular edges.

(ii)   Types. The following stones have been commonly used in the past and are still exported (albeit in rather smaller quantities) to parts of the country distant from the quarries. Their high price is due both to shortness of supply and transport costs. Other stones have been used locally but it might now prove difficult to find a quarry still working its stone in the way which gives flags. Second-hand flags may be found; indeed there are a number of companies about the country dealing exclusively in such salvage.

York stone is a sandstone. It is probably the best known of all paving stones. Its colours are brown through to buff. Its geological history allows it to be split to appropriate thicknesses, giving a riven face. A rubbed finish that is less irregular is also available at a higher price. Thicknesses (particularly when with a riven face) will vary between 50 and 63 mm (2–2.5″) within any one load and the way in which it is used must allow for that.

Portland stone is an oolitic limestone. Whilst in the Cotswolds, for example, oolitic limestones can be split to a riven face, the Portland beds have much greater consistency. For this reason it is usually sawn to thickness giving a characteristic texture, even thickness and a high price. Again a rubbed finish removes the sawn texture. This stone is quite soft and should not be used where heavy wear is expected. Its colour is white or very pale grey, but when used for paving flags, brightness soon fades. Slab thickness may vary between loads from 50 to 63 mm (2–2.5").

Purbeck Portland is harder than ordinary Portland stone, making it even more expensive. In other respects it is the same.

Slate is a metamorphic rock formed deep underground when sediments were heated by adjacent volcanic activity. Its crystalline structure was conditioned by the pressure of that circumstance, giving cleavage planes on which the stone can easily be split. It can give one of the hardest wearing stones for paving use – provided that it does not have too many cleavage planes, for these can be opened by frost. Natural riven, frame-sawn, sanded, and finely rubbed finishes are available, in that order of increasing cost. The smoother finishes might prove slippery in exposed areas. Welsh slate is blue-black to blue-purple whilst Lake District and Cornish quarries produce a variety of greens, some with brown markings. Slate flags can be thinner than other stones, 25–32 mm being common and 15 mm being possible with domestic loads on a very even and well consolidated base, or on *in situ* concrete slab.

Bath stone is another oolitic limestone. It would usually be considered too soft for paving flags, but it can have a value in areas with only very light foot traffic, where irregular and worn qualities might be in character. Finishes as for Portland stone with thicknesses which tend to be slightly more than given for that stone.

(iii) Patterns of laying. As already stated natural stone is an expensive material. Quarrymen will, then, split or cut the flags in ways which waste as little as possible. This leads to a wide range of sizes up to a maximum. Whilst quarries will provide flags of consistent size and (as far as proves possible) thickness, they will charge as much for the material wasted as for that despatched. Hence stone flags of a regular size should only be considered for the most prestigious of projects.

A range of consistent widths with random lengths is more commonly available, particularly with slates. A simple modulated pattern can be created with such slabs, with a value in both formal and informal contexts. Figure 1.18 illustrates two possibilities. But it should be remembered that the widths of the material being quarried changes. Potential suppliers should be contacted for up-to-date information, but even then it would be best to design with some flexibility so that an idea is not totally invalidated at a later date.

Finally, we would like to make a plea on behalf of the much maligned crazy paving. Admittedly an over-dominant area using broken pimply concrete flags can look pretty awful. But in the hands of an imaginative designer, stone can be used in this way to great effect, perhaps when the cost of stone could not be justified if used in any other way.

(iv) Method of laying. Other than the implied need to take greater

73

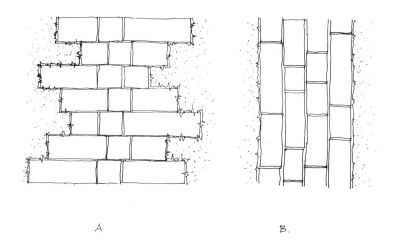

A   B.

1.18 *Two ways in which stone flags of consistent width but random length might be laid with relative formal or informal effects.*

care, stone flags are laid in the same way as described in section 1.15.3.

1.15.3   Precast concrete flags

(i)   Uses. The precast concrete slab is ubiquitous since it is now used wherever and whenever natural stone would be too expensive. Which is not to say that these flags can be used as a direct substitute for the natural material. Their qualities are not identical, even in those cases where it was clearly the manufacturer's intention to make them so. The machine-made precision of the precast slab is the dominant quality in any area paved with that material. This is a quality that an area of stone flags will never have, even when individually dressed with machine-like precision. Designers should always understand and exploit the potentials of their materials. Precast concrete flag pavings should be designed in the knowledge of what they can be, not in the forlorn hope that they will look like something else.

The smaller thicker concrete flags, when consistently bedded on a consistent base, would be able to take light and occasional vehicular traffic. But their lack of tensile strength makes them very vulnerable to cracking should loads be high and/or there be any irregularity in their support. For this reason they cannot be recommended as a surface for anywhere other than pedestrian-only areas.

(ii)   Types and sources. These are many and various – too many and too various to attempt to describe them all. A very broad sub-division might identify two categories, plain and textured. A further sub-division of the latter might distinguish those textured through moulding and those textured by exposing the aggregate. An alternative very broad sub-division might identify plain and coloured, with the latter being further split into coloured by adding pigment and coloured by

exposing the aggregate. BS368:1970: 'Precast concrete flags' concentrates on the basic product with a limited range of finishes and sizes. However, most manufacturers also follow the standard in so far as it is applicable to their more decorative products.

Designers should be aware of the ranges of products available and then seek specific information from manufacturers' catalogues. The National Paving and Kerb Association will help to direct enquiries to possible suppliers.

The basic product is made in a hydraulic press exerting a pressure in excess of 7 N/mm^2 (1000 lb/in^2). The conditions of manufacture make it both difficult and expensive to alter the standard mould. These flags are self-coloured (neutral grey), smooth on one side (usually used uppermost) and with pimples on the other which are a result of the manufacturing process and are not added solely as decoration.

BS368:1971 includes one width, 600 mm (2'0"), four lengths, 900, 750, 600, and 450 mm (3'0", 2'6", 2'0", and 1'6"), and two thicknesses, 63 and 50 mm (2'6" and 2").

The Standard also covers the cements and aggregates, consistency, and freedom from defects such as irregular thicknesses and angles, malformed edges, and untrue or warped surfaces. There can be no doubt that a BS precast concrete flag has a high technical quality which, when correctly laid, gives a durable if somewhat utilitarian pavement.

It is in order to give a product with comparable technical qualities but with rather more aesthetic potential, that non-standard flags are made. These tend to be smaller and thinner than a BS flag, and the majority have a finish other than plain and self-coloured. Thicknesses vary between 38 and 50 mm (1.5–2") and sometimes even thinner. Ratios of side lengths tend to be 1:1, 1:1.5, or 1:2, but within those limits close to every permutation is available from 113 mm (4.5") through 150 mm (6"), 225 mm (9"), 300 mm (1'0"), 450 mm (1'6") and 500 mm (1'8"), to 600 mm (2'0"). Further, there are flags made which are non-rectangular; some, such as octagons, offered for use with rectangular or square flags; some, such as hexagons, to give other all-over patterns; and some, such as circles, for alternative uses such as stepping stones in grass. These smaller units, even when plain and self-coloured, have much more potential when scale needs to be reduced and/or visual interest increased, than have the standard forms. However, some care needs to be taken when considering using coloured or textured flags – there can be temptation to create complex and polychromatic patterns that look very exciting on the drawing board but which might be disturbing underfoot, other than in very active places. Figure 1.19 illustrates some patterns possible with non-rectangular flags.

(iii) *Pigmented flags.* Pigments can be added to precast concrete flags in exactly the way that they can be used to colour *in situ* work. The pigment might be through the whole of the flag or in only its

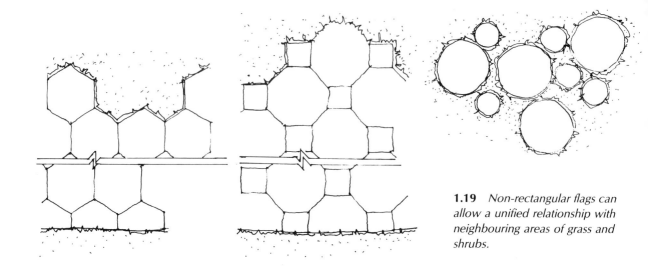

1.19 *Non-rectangular flags can allow a unified relationship with neighbouring areas of grass and shrubs.*

upper surface. In the latter case that thickness should be at least 13 mm (0.5″) to ensure that there is no possible risk of patchiness through wear. There is a tendency for chroma to fade with time although the problem of patchiness through pigments washing out has now been largely overcome. Shades vary from blue-black, through browns and brown-reds, to buffs, yellows and pinks.

(iv)   Textured flags – moulded. Early moulded flags used rubber mats which were generally produced for other purposes (bath mats and car mats, as examples) to line the mould in which the flags were cast. Considerably more sophistication is now employed. The moulds are purpose-made either to give abstract pattern or, more frequently, to simulate the surface of natural stone flags. The more sophisticated make their moulds from original examples, and the most sophisticated use a large number of originals and mix the resulting flags within any one delivery. These are then irregularly pigmented and can, as individuals, be virtually indistinguishable from the original. But, as has been said, their regularity of shape still characterizes an area of these flags. The paradox remains that moulded flags which hint at, rather than ape, the natural material, are frequently the more satisfactory.

Other flags are made in steel rather than rubber lined moulds. These tend to give bold patterns rather than textured surfaces. A few examples of many might be; 600 × 600 mm (2′0″ × 2′0″) flags with grooves which sub-divide its surface to give four 150 × 600 mm (6″ × 2′0″) bars, or sixteen 150 × 150 mm (6″ × 6″) squares; flags which are grooved so closely as to give a rilled effect; and others covered with pyramids with significant height. These are offered because of their combined functional and visual qualities. The grooves give more grip whilst reducing scale, and because the patterns are bold the larger aggregates are not excluded from the surface, giving harder wearing flags than those with a finer moulded texture. Those with large pyramids (or similar) can give a visually clear line (say at a road kerb), at places where pedestrians are to be deterred. Whilst there is no technical reason for this, the majority of steel-moulded flags are given darker-toned colours, if they are coloured at all. Figure 1.20 shows some of the available steel moulded flags.

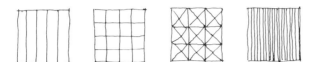

1.20 *Precast concrete flags can be obtained with a range of cast textures.*

(v) *Textured flags – exposed aggregate.* The aggregates in precast concrete flags can be exposed in the same ways as can those in *in situ* slabs. But, because all of the work can be undertaken in a factory, a higher consistency can be achieved than is, on average, the case with site work.

The textural effect created will be dependent upon the aggregates used, the technique with which, and the depth to which, they are exposed and, to a lesser extent, whether or not coloured cement is used.

The aggregate should not be very rounded in form or it may pluck out of the surface when exposed, and it would not provide a good grip. For this second reason, aggregates which have or would wear to, a smooth polished finish are not appropriate. But very sharp aggregate, whilst excellent in those respects, should also be avoided in case of an accident. A large aggregate size of 19 mm (¾″) is a maximum with hydraulically precast flags, for technical reasons. Further, that is the largest size that is likely to give appropriate scale in pedestrian areas. Aggregates of that size do not have to be used if a finer texture is desired, the practical minimum for the largest size being about 10 mm (⅜″).

Colours are limited only by those naturally available, and overall effects that are never seen in nature can be created by mixing different aggregates in the same slab.

Three techniques for exposing aggregates might be used. First, the flag is brushed and gently washed very soon after casting. This takes away the cement and sands in the upper surface leaving the larger aggregates exposed. Secondly, the flag can be wire-brushed or grit-blasted after it has hardened. This is a more expensive process than the first but gives a more angular texture. Thirdly, the aggregates to be seen can be scattered over the surface of the flag shortly after casting, and tamped in. This last technique, whilst economical with expensive aggregates, carries the risk that aggregates will pluck out in use.

A fourth technique should be mentioned, although it only has limited application. Hand-cast flags might be made (usually rather larger and certainly rather thicker than the norm) and relatively larger aggregate (such as small cobbles) hand placed into the wet surface. An almost identical effect can be created on site working directly into a wet *in situ* slab. The relative advantages of precast and *in situ* work would have to be assessed in each specific case.

Significantly different effects can be achieved by modifying nothing other than the depth of exposure. With increasing depth more of the form of the aggregates will be seen, and less of the cement and sand matrix; the texture will become coarser and the colour increasingly determined by the larger aggregate. But care must be taken not to expose too deeply or the aggregate may come loose.

Pigments can be added to the cement in the same range of colours as is available for untextured flags and *in situ* concrete. This, however, does not have a significant effect as much of the visible cement is taken away when aggregate is exposed. Gap-graded aggregates might be used which will then leave more of the matrix exposed as a background to the larger aggregates.

It should also be remembered that the colour of the sand used in any precast flag will affect the surface colour. This is so whether or not aggregates are to be exposed – indeed the process by which surface laitance is worn away by use and weather could be added as a fifth technique for exposing aggregate, but in this case only the fine aggregate. A gap-graded mix, where care has been taken to select a sand of appropriate colour, is more likely to give a product with genuine quality, than would be achieved by adding artificial pigment.

It should be noted that textured flags, particularly fine textured flags, will 'weather' both more quickly and more consistently than will smooth. Moreover, should they become accidentally stained, that staining is likely to be less of a disfigurement.

(vi) Laying precast concrete flags. A formation must be prepared and a base laid as described in sections 1.6.5. and 1.7.4. Cross-falls of 1:40 and longitudinal falls of 1:120, or thereabouts, should be established when the formation is constructed. A 75 mm (3") deep base is likely to suffice in most circumstances and rarely would more than 100 mm (4") be needed.

Four related but different approaches are then possible. Each has advantages and disadvantages and so which is to be specified should be judged in the light of specifics.

(a) The base is blinded with a 38 to 50 mm (1.5–2") deep bed of clean and reasonably dry sand. (On a very true and well consolidated base, this depth might be reduced to 25 mm (1"). The sand should be sharp and well graded from, say 3 mm down to 0.25 mm (0.12"–0.01"). It should be screeded to level, although this need not be done as precisely as is necessary for interlocking blocks. The flags are then laid on the sand as far as possible in the precise place in which they are to remain. Care must be taken when the flag is lifted into place to ensure that the sand bed is not seriously distorted, by allowing a corner to drop, for example. Simple tools are available to help with this.

A large, long handled wooden or rubber mallet is used to adjust position and level, but too much cannot be done about this without breaking the flag. If it is too high or too low it must be lifted again and the sand bed adjusted.

(b) The procedure described above is exactly duplicated except that a dry mix of cement and sand, lime and sand, or cement, lime and sand is used for the bed.

A proportion of 1:5 of cement or lime to sand is usual; or say 1:1:10 of cement, lime, and sand. Being laid dry, this approach has all the advantages of a sand bed and, in addition, the cement and/or lime will take up water from the air or ground and achieve a set. This removes any risk of the sand escaping from the bed due to erosion by rain water, defective edgings, or whatever.

(c) The cement:sand, lime:sand or cement:lime:sand bed can be laid wet. Similar proportions of the materials as described above, are mixed with a minimum of water. The base is given a sand blinding to fill surface irregularities and a full bed of mortar spread over an area that is a little larger than the next flag to be laid. The principal advantage of this technique over (b) is that the mortar can be given exactly the amount of water necessary for it to achieve all of its potential strength, which makes this a more satisfactory approach if occasional light vehicular traffic is antici- pated. But it does have a disadvantage. Any sand or dry mix which gets on to the surface of a flag can simply be brushed off. With a wet mix this is not possible, leading to a disfigured surface – even when dilute hydrochloric acid is used as a part of the cleaning operation.

(d) Less mortar can be used, and under greater control, if it is placed on the blinded base in blobs rather than as a full bed. The mortar must be mixed stiff so that those blobs can hold the weight of the flag at the correct level, until they have set; but easing the flag down to its correct level is rather easier than with a full bed, be that wet or dry. There is a real disadvantage to this technique which makes it inappropriate for any other than the most lightly loaded of circumstances. Each flag is left bridging between the mortar spots. Concrete has low tensile strength and any bridge experiences considerable tensile stresses. Whilst the problem can be minimized by making the blobs of decent size and placing them so that bridge distances are small, the fact remains that flags laid using this approach are much more likely to crack in use than with any other. Figure 1.21 illustrates how a flag laid on mortar blobs compares with one laid on a full bed.

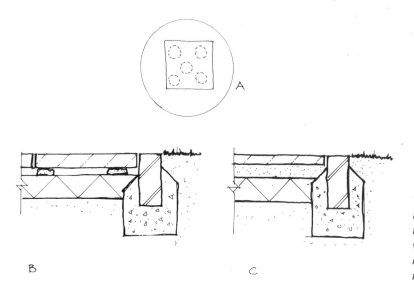

B

C

1.21 *A and B show paving flags laid on mortar blobs. This technique should not be used where flags are thin and/or loads more than minimum. C shows flags laid upon a full bed.*

Whatever bed might be used, joints can be butted or open. Butt joints reduce the amount of labour needed for the pavement as a whole, and decrease the likelihood of grass and/or moss establishing. But no butt joints fit entirely flush and the surface should be swept over with dry sand or dry mix, working it into the joints with a stiff brush.

Open joints might be used when a pattern in the flags is to be stressed, where grass and moss is to be encouraged, or where the open joint is intended to assist drainage (but that is not recommended). In these cases the joint is usually about 10 mm (⅜″) or a little more. The joints should be filled with sand or fine stone chippings, or a mixture of sand and topsoil, or a dry mix of cement and/or lime and sand, all depending upon the objective of laying with open joints. Yet wider joints might be included, possibly of varying widths, where grass, mosses or other vegetation is to be an integral part of the design. This kind of detail should be considered in the light of that which follows.

(vii) Flags laid in grass or gravel. Stone or precast flags let into grass as stepping stones can be a most attractive way of reinforcing that surface whilst remaining essentially casual, or to assist with a transition between formal and casual areas. Flags might also be let into a loose gravel area either to give an easier surface underfoot or as a way mark. In both cases the flags must be placed at centres which correspond to pace length, to ensure that feet land comfortably on the flags rather than the wide joints between. Those centres will, then, vary depending upon the size of the flag used. Average pace length remains constant at about 750 mm (2′6″) although pedestrian speed must also be anticipated – a stroll uses much shorter paces than a walk.

If the grass already exists, the flags should be placed (and possibly adjusted to achieve precisely the desired effect) and a sharp spade used to cut down through the turf exactly on the edges of each flag. Flags then have to be lifted aside and the turf removed, together with any underlying topsoil to a depth of flag thickness plus 50 mm (2″). That formation should then be carefully levelled and a sand bed of 50 mm plus (2″ plus) spread and screeded. The flag can then be replaced and gently tamped to level.

If the grass does not exist, or if a greater weight of pedestrian traffic is anticipated than the above technique would satisfy, or if the flags are to be placed in gravel, a base should also be incorporated (see section 1.7.4). This is unlikely to have to be more than 75–100 mm (3–4″) thick. If the flags are to be laid in gravel, it will be part of the general base. If the flags are to be laid in grass or other ground-cover vegetation, it should be a strip, not individual patches, running along the general line that the stepping stones are to take and extending beyond their width by 200 to 300 mm (8–12″). In the latter case, the base should be consolidated to no greater degree than necessary to support the loads, which will not be high, otherwise problems of drainage might result. The flags should then be laid using the wet-mixed, full-bed technique; dry materials will never be adequately contained at the edges of the flags. Figure 1.22 illustrates this technique.

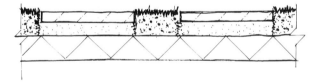

1.22 *When flags are laid as stepping stones within grass, it may be best to lay the whole of the path on a hardcore base to ensure drainage. The grass between the flags is then seeded into a 50/50 topsoil/gravel mix.*

A minimum overall depth for flag plus bed of about 75 mm (3″) will be needed if the spaces between and around the flags are to have an adequate depth for the soil and this might have to be more if thick flags are used, or the base is very irregular. If loose gravel is to be the background, this can be simply spread and raked to levels. For grass and other ground covers it is best to spread not pure topsoil but a mixture of topsoil and small aggregate. A proportion of 40:60 aggregate:topsoil should be satisfactory. The aggregate size can be quite small, say 5 mm (0.2″) single sized. That aggregate ensures free drainage through the topsoil and gives a little more for the plant roots to take hold of, and so be a little less vulnerable to trampling. (Indeed, grass areas which take occasional foot traffic along a clearly defined line, can be reinforced in this way to some effect. This detail is, perhaps, the simplest possible form for a pavement.)

It should be added that when flags are to be introduced into existing grass, they might be cast *in situ* directly into the mould made by removing a 75 mm (3″) or so, thick turf. There are then no constraints on shape or size. If an exposed aggregate finish is considered then it would be best to tamp this into the surface rather than to try to expose that which is already in the slab. Otherwise the surrounding grass will get in a terrible mess.

(viii) Laying flags on an in situ *concrete base.* Thick stone or precast flags laid on a wet-mix bed on a substantial and well consolidated base, would be able to take occasional light vehicular traffic. But that detail could not be recommended if either weight or density of traffic were higher. If an extensive area of pavement were to be made available to heavy traffic, then flags would not be a recommended surface finish. But it can happen that a proportion of a larger pedestrian space has relatively infrequently to take motor vehicles (weekly visits for refuse collection, perhaps). Given that a change of surface was not to be incorporated as a signal of possible danger, or if that signal need not be over-dominant, then there might be value in maintaining the original flagged surface finish, but over an *in situ* concrete base.

The base and its sub-base should be designed and constructed as described in section 1.13. It would be constructionally convenient to set the surface of the *in situ* base at the same level as that of the regular base used generally. If possible, a generous width of base should be provided so as to avoid any risk of the vehicles running off on to the surrounding flags. (Alternatively, bollards can be used to restrict vehicular movement.) The flags can be laid through at the same time as generally, but whatever bed they have elsewhere, they must have a full, wet-mixed bed over the *in situ* slab. No other bed would guarantee that the loads of the vehicles are transferred through the flags and their bed, to the base. If those loads are not successfully transferred, the flags will crack.

1.15.4 Brick pavings

(i) Uses. Until the advent of vibrated block pavements, brick-sized elements were not greatly used for road construction in Britain. This situation has now changed and in some kinds of places, such as housing estates, small scale, patterned surfaces are becoming the norm. But it was not that traditionally constructed brick pavings were inappropriate for road construction. In Holland, for example, millions of bricks were and continue to be used for this purpose every year; bedded directly on sand when the vehicles are slow moving and on wet mortar for quicker and/or denser traffic. But that is now largely by-the-by. Interlocking block pavements (described in section 1.14) provide excellent surfaces for vehicular traffic. In this section, then,

Satisfying function (however prosaic) can also
contribute to character.

attention will be concentrated on traditional brick pavings for pedestrian use.

(ii) Characteristics. The building brick has developed over the centuries to have a size and weight that is convenient for building brick walls. These dimensions are not so convenient when building brick pavements. Further, the fact that the bed face (that which is 215 × 103 mm (8½ × 4″) or thereabouts) is rarely seen in wall construction means that it is frequently not fit to be seen – yet it is that which provides the pavement surface when the brick is used most economically as a paver. And further, bricks used as pavers are likely to be more thoroughly wetted, and for longer periods, than when used in a wall, and so may need qualities of frost and sulphate resistance that would be unnecessary for wall construction.

All this must lead to the conclusion that bricks made specifically as pavers are always likely to be technically and economically preferable. However, the building brick can sometimes be visually softer than a paver, and sometimes there can be a need to unify what might otherwise have been too diverse a design, by using the same material for floorscape and walls. In such circumstances the building brick might well be considered as a paver, but must then be selected with some care.

The selected brick must be capable of transferring imposed loads to the pavement base; must be resistant to wear; must provide a non-slip surface; must be able to resist sulphate attack; must not release sulphates if used with cement mortar; and must not be adversely affected by frost – all in addition to having appropriate aesthetic potential.

The first of these presents little difficulty. It is most unlikely that a brick with the strength necessary in a wall would not also have the strength necessary in a lightly loaded pavement for, as with other flexible structures, it will be the base which carries the bulk of the load. So, provided that the base and the bed are soundly constructed, the pavement is unlikely to fail in this respect.

There is rather less certainty about resistance to wear. Even if the designer could quantify predicted wear, it is improbable that any manufacturer could provide that kind of data for non-pavers, unless, that is, there is experience of that brick used as a paver. All that can be said is that bricks with high compression strength (and those data will be available) are likely to wear well, remembering, of course, that bricks used in the circumstances being considered here may not be subjected to a great deal of wear, and could possibly look better if they were.

The non-slip qualities of a brick should be considered when it is dry, when it is wet, and when it is frosty. In these first two respects, the matter can be relatively easily judged. If the face of the brick that is to be used as the paving surface has a texture then it will be non-slip both when wet and dry. But this should not be taken to mean that a very smooth brick would be inappropriate. Indeed, many of the bricks made specifically as pavers have smooth surfaces. However, if a smooth building brick is being considered as a paver then it would be advisable to consider the way in which it is to be laid in terms of practicalities as well as visual potentials. The joints between the bricks can provide necessary grip. A decision might, then, be taken to use a very smooth brick so that it is the stretcher face (that which is 215 × 65 mm (8½ × 2½") or thereabouts) which shows, and with open rather than butt joints. Such an approach would give the maximum number of edges to provide grip. (But it would also, unfortunately, add considerably to cost.)

There is more of a dilemma when considering the slipperiness of bricks in frosty weather. There is a tendency for those bricks which are most resistant to frost, to extrude ice from their microscopic pores for some time after freezing conditions have passed. This can be particularly dangerous, for users may not be aware that there is a continuing need to take care. Again, the difficulty can be reduced by using the bricks in the way described immediately above, and by working closely with the manufacturer when selecting a brick for a particular locality.

Certain salts (magnesium and calcium sulphates in particular) can combine chemically with constituents of cement, causing both an expansion and softening of the mortar or concrete matrix of which that cement is a part. Both the stability and the durability of brickwork might be affected in this way. With properly designed and constructed brick walls (whatever material the bricks are made of) this is a rare problem, for the brickwork must be saturated if the salts are to be liberated in sufficient quantities to cause a difficulty. Walls are only likely to be saturated below faulty copings or sills, as examples. But bricks used as pavers are very likely to be saturated for significant periods. So, if there are any sulphates present in the ground, concrete bricks should not be used unless the manufacturer has used sulphate-resisting cement. Similarly, if the bed and/or joints are to include cement, it should be sulphate-resisting, whatever the material of the bricks. But the sulphates may be present in the bricks. In this case the pavement might be laid without using cement, or using only sulphate-resisting cement, or the bricks might be selected because they have a low sulphate content.

The Brick Development Association (BDA) have said that 'there is no laboratory test which will satisfactorily predict the frost resistance of a brick'. Whilst it is known that most bricks with high compression strengths and low water absorption rates are frost resistant, it is also known that there are some with low compressive strengths and very high water absorption rates that are also frost resistant. A speculative explanation of this paradox seems to be that both high compressive strength and low water absorption are related to density whilst frost resistance is dependent upon cell-wall strength, and upon the brick being well fired. So it is that a very light, porous brick might be frost resistant if the material that is there is very strong. The BDA suggest that there is no real substitute for direct experience (either by the designer or manufacturer) when judging the frost resistance of a particular brick in particular degrees of exposure.

Indeed, it cannot be over-emphasized that that advice applies when judging all the characteristics of a brick which is not specifically manufactured as a paver but which is, nevertheless, to be used for that purpose.

(iii) Patterns of laying (bonds). Bonding brick pavings for pedestrian use is of significantly less technical importance than it is with roads (see section 1.14.5). The most energetic pedestrians are not likely to physically disturb bricks as they walk and turn. The decision as to the pattern to be used is, then, most likely to be judged against aesthetic criteria. Figure 1.23 illustrates examples. But it must be remembered that the size of building bricks is conditioned by their convenient use in a wall. Basket weave, for example, can only be laid with open joints, as illustrated.

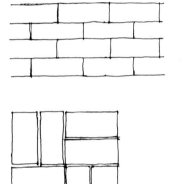

A

(iv) Drainage. Relatively narrow paths getting very little use, and in the most casual of contexts, might be laid with generous sand-filled open joints over a sand bed on a lightly consolidated base – all with a view to allowing rainwater simply to percolate through the surface and into the ground. Even then, this approach would only be advisable where short term flooding and longer term deterioration would not be unacceptable. Generally speaking, brick pavings should be laid with 1 in 40 (2.5%) cross-falls and 1 in 80 to 1 in 120 (approx. 1%) longitudinal falls. Drainage might be to running-bond channels leading to gulleys, or to linear drains, or simply to a gravel-filled ditch with or without a land drain. But, whatever option is adopted it must be considered as part and parcel of the whole design and not left until after bricks have been selected and/or bonding pattern agreed.

B

(v) Construction. Traditional brick pavings are essentially constructed in the same ways as are stone or precast concrete flags. The area in which the pavement is to be built must be excavated to an appropriate depth (the depth of the base, plus depth of bed, plus depth of brick) and a formation prepared that is free of deleterious material and with the falls that are to be given to the pavement surface. The base of hardcore, broken stone or other suitable material must then be compacted sufficiently to ensure that there will be no settlement in use. 75 to 100 mm (3–4"), consolidated with a light roller or its equivalent, will be sufficient in the vast majority of cases.

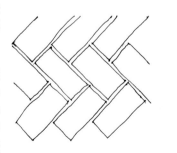

C

1.23 *Paving bricks can be laid in a range of different bonds, of which these are but three. A is running bond, B is basket weave and C is herringbone.*

A bedding depth of 38 to 50 mm (1.5–2") is needed. The bed can be dry sand, a sand:cement or sand:lime:cement mix laid dry, or a similar mix laid wet. The bricks are laid individually and lightly tapped or worked down by hand to level. It must be remembered that if a dry bed is used the edgings must have been designed and constructed with some care, and built before the brick paving is laid, to ensure that the bed is securely contained. Figure 1.24 illustrates these points.

If laid butt jointed, the surface should be brushed over with dry sand using a stiff-bristled broom, particularly over the joints and particularly if the bricks have irregularities giving uneven widths to the joints. If laid open jointed, the joints can be filled with dry sand, dry mix, wet mix, topsoil, or a top soil:sand mixture. It would be a mistake to fill the joints with a material with more potential strength than had the bed (dry mix over a sand bed, or wet mix over a dry mix bed, for example). But filling the joints with a weaker material might well be considered, particularly if this avoided the possibility of staining the surface with cement. Soil (in whole or in part) would only be included in the joint if it was intended that mosses or other vegetation were to be encouraged. This should only be done in the knowledge that the surface will be more slippery and the bricks more vulnerable to attack by frost, than would otherwise be the case.

The colour of open joints that are to be kept free of vegetation will be of great importance. In such circumstances cement or cement and lime will have to be included with sand to fill the joints; both to discourage the vegetation in the first place and to make it easier to clear should it manage to gain a toe-hold. The mortar can be coloured by adding a pigment but it is likely that it can be adjusted purely through careful selection of sand. Experimental panels should be made and that which is most successful retained to record the standard to be achieved with the work proper. Given that this would not create practical difficulties, these panels might also experiment with recessed joints of different depths. It could be that shadow lines rather than coloured mortar is all that is needed.

(vi) Traditional brick pavings for occasional vehicular traffic. As with precast concrete and stone flags, it cannot be recommended that traditional brick pavings be laid for regular vehicular traffic unless the circumstances are very special or if the traffic will use only a small part of a much larger area.

Bricks of high compressive strength, laid to show their stretcher face laid on a cement:sand or cement:lime:sand bed with a strength that is compatible with that of the bricks (their manufacturer will advise), over a well consolidated base designed as described in section 1.7.4, could take regular and heavy vehicular loads, provided that speeds were not excessive.

It is not, then, practicalities which limit the use of bricks in this way; it is economics. Very special circumstances can be accommodated if the budget is adequate.

Those parts of larger areas that will occasionally be crossed by vehicles could also be treated in this way. Indeed, there could be an advantage in this if, for example, the finer texture given by the

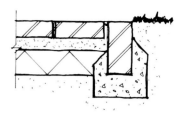

1.24 *Areas of paving bricks must be restrained at their edges for the sake of both constructional validity and appearance.*

stretcher faces were used as a warning signal within an area which, generally, showed the bed faces of the bricks. However, there could again be severe cost penalties if no such differentiation was wanted. Either something like 50% more bricks would be needed (if all were to show their stretcher face) or individually more expensive bricks would have to be bought (if all were to have the strength that would be required only where the traffic is to cross).

Under such circumstances it would be better to consider the bricks as a surface course over the rigid *in situ* concrete road, or over a flexible lean-mix concrete base. Whatever the base, it should be designed as described in sections 1.7.4 or 1.13, and the bricks bedded on a wet mix with proportions of sand and cement or sand, cement and lime, as advised by the brick manufacturer. Open joints would have to be used (filled, wet or dry with a similar mix to the bed) to ensure firm adhesion, and this would dictate the jointing to be used elsewhere.

So, it must be said again that it is only in the most special of circumstances that traditionally laid brick pavings could be seen to have advantages over bricks/blocks made specifically as pavers. With those, where additional strength is needed, the depth of base can easily be increased and/or thicker elements (of the same material) laid, with the whole of the surface bonded and bedded as one.

(vii) Traditional brick pavers. We have to go back through several technological revolutions to reach a time at which engineering brick pavers were commonly employed in stable yards and similar locations, both domestic and commercial. A number of companies continue to manufacture these to the original nineteenth century patterns, and firms are establishing to deal in these, and other, salvaged materials.

Clay engineering bricks are of high density, low porosity, and high compressive strength. Their colour varies depending upon the geological deposits and the firing process used. Blue-black and then red are the commonest colours and a smaller number of buffs are also made. First quality pavers will be of a consistent colour but second quality will show colour variations between individuals and, sometimes, within an individual brick. Seconds are cheaper than firsts although it will be interesting to see whether this remains the case as these pavers are increasingly used for their decorative rather than utilitarian qualities.

Engineering pavers are very well fired, the aim being to produce a high strength vitrified body. This also gives a very slippery surface. So, whilst plain pavers are available they are more generally produced with chamfered edges or with a square or diamond chequered pattern; all with the object of providing grip. Some of these traditional patterns are shown in Figure 1.25. They are laid in the way described above. However, because they are precisely made they are usually laid with a thin open joint of only about 3 mm (⅛"). But the traditional 9 × 4.5" (225 × 112.5 mm) format is maintained giving bed

1.25 *Traditionally, stable pavers in engineering or semi-engineering bricks were available either with a smooth finish or a range of textures. Items such as these are now more likely to be found as salvage than obtained new.*

faces of 222 × 110 mm or thereabouts. Thicknesses vary between 38 and 72 mm (1½ – 3″) with 50 mm (2″) being the commonest.

The points which were made above about drainage are of particular importance here. The more precisely the unit is made the more precisely it must be used.

1.15.5 Sett pavings

(i) The materials. Setts have a smaller plan area than do flags but are usually considerably thicker. That being so, their form is essentially cuboid. The bulk of this section will consider stone setts, but there are also alternatives which are discussed at the end.

Stone setts are made from both igneous and sedimentary rocks. Because of their lower strengths, gritstone and similar sedimentary setts tend to be larger. As far as we have been able to ascertain, sedimentary stone setts are no longer manufactured although it might be possible to have them made to special order – at phenomenal cost. Moreover, the second-hand supply, which was once quite good as older streets were taken up and relaid with tarmacadam, is now virtually exhausted.

Igneous rocks with a value to the building industry are all known as granites to that trade. Whilst a geologist would demur, it seems sensible to adopt that term. The second-hand supply of granite setts is much the same as with gritstones and similar. But granite setts are still made and there is a current standard specification (BS435:1975). Their depths range between 75 and 125 mm (3–5″), their widths between 100 and 150 mm (4–6″) and their lengths between 100 and 250 mm (4–10″). All combinations of these sizes are not commonly available, the longer lengths, for example, being generally combined with the greater depths and breadths. Granite setts are not cut to the sizes required, but broken with an edged mechanical hammer. They cannot, therefore, be entirely regular (a quality which adds much to their aesthetic potential). There is a tendency for them to taper within their depth, but this is limited by the BS, as are their other possible irregularities.

In theory, at least, their colours can vary over the whole spectrum of naturally occurring granites. However, for reasons which seem to be largely a matter of economic coincidence, the majority of new granite setts are grey with, sometimes, black flecks, whilst second-hand setts more often include reds and red-browns, and other darker tones. It should be added that this impression will, in part, be conditioned by weathering and use. A second-hand sett, with a century or more of hard use, will inevitably be darker toned than newly exposed stone. And use also affects regularity. Second-hand setts often have more of the qualities of a squared cobble than the crisp irregularity of the new material.

(ii) Uses. Stone setts were widely used to pave roads and squares

at a time when labour costs were low. Setts could then be both manufactured and placed very cheaply. This is no longer the case, certainly in Britain where the major centres of population are, in the main, a considerable distance from the granite outcrops. Moreover, rubber tyres are much less hard on a road surface than were steel shod wheels and hoofs, but they do require other qualities of it if they are to maintain grip. Whether it is reducing demand that has pushed up the price of setts, or vice-versa, is something of an academic question. The fact remains that it is now only rarely that the cost of paving extensive areas with setts can be justified.

Much more common is the use of setts with another material. They have a particular value in that their high compressive strength means that they can take the weight of a roller (given, of course, an adequate foundation). So, setts can be used to give a smaller more intimate scale to a surface that is generally paved with tarmac or bitmac. The range of sizes they come in allows, for example, three or four rows to be used to form a channel (quite possibly dished) with a width that precisely matches the gully grating to which they lead. Similarly, they might be used to permanently mark the bays in a car park (much to be preferred to grubby white lines), or through both their visual and physical qualities offer a signal to drivers that they are moving off a surface that is predominantly theirs on to one to which pedestrians have at least equal claim. And, as a final example, they can be used to aid the visual transition between areas of different and contrasting materials. Figure 1.26 shows setts used in combination with tar/bitmac.

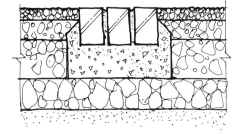

1.26 *Stone setts can be used with effect as trim, which enlivens bound macadam pavements. For example, the detail shown here might be used to mark the spaces within a car-park.*

(iii) *Construction.* As with all other pavings, setts must be laid on a suitable foundation. The formation must be designed and prepared (after due consideration of falls, drainage, etc.) as described in section 1.7.3 and on this must be consolidated a base (and possibly sub-base) as described in section 1.7.4. Overall depths should be designed so that the upper surface of the base is the thickness of the setts to be used plus 25 mm (1″) below the finished surface.

The setts are laid on a 25 mm (1″) bed of sand, or 1:3 cement:sand dry mix, and tapped down to level to leave an approximate 10 mm joint. Two bonds are common (the second less common in Britain than in other parts of Europe). Running or breaking bond constitutes the majority of traditional British work. Elsewhere, a fan pattern is more usual. Arcs of setts are laid using a one-legged stool on which the layer pivots, with the radius of the arc being determined by arm length. The stool is shuffled back by 100 mm (4″) or so after each arc is completed, with the pavement developing in strips about 1.5 m (5′0″) in width. The nature of the process leaves a chevroned edge to each strip from which the arcs of the next strip are generated. Figure 1.27 shows the appearance of these two bonds.

When bedded on sand (the traditional detail) the joints are filled to within 25 mm (1″) of the surface with coarse sand/fine grit and finished with a cement:sand grout. When bedded on a dry mix they can be brushed over with more of the same, either to finished level (if the pavement is not to be used immediately) or so as to take a wet grout as before. In either case it is important that the surface is sealed before being exposed to heavy rain which would be liable to wash out the bed.

(iv) Alternatives to stone setts. Some precast concrete manufacturers make products that can be considered as an alternative to granite setts. These are moulded in steel to a range of sizes comparable to the smaller dimensions given for granite setts. Excess concrete is taken off of the surface of the moulds with a mechanical broom. This gives a characteristic texture which evokes that of the split stone without attempting replication; an excellent example of concrete being used in a way which exploits its own potentials, rather than requiring it to pretend to be something else. They can be used in exactly the same ways as described above.

Other manufacturers offer interlocking concrete blocks with forms akin to those of worn sedimentary setts. These are vibrated into a sand bed as described in section 1.14.

1.15.6 Cobble pavements

(i) The materials. Cobbles are fist-sized and egg-shaped pieces of hard igneous and metamorphic rocks that have been given that form by being tumbled in streams, rivers, and the sea. Whilst it is illegal simply to carry off materials from beaches or river beds, there are firms with appropriate licences from which cobbles can be purchased in either single or mixed sizes. Further, strata of cobbles are sometimes found by firms quarrying sedimentary rocks for other purposes, and in sufficient quantities for them to separate these for sale. Beds of water-worn flints found in chalk being quarried for cement is an example. Finally, there is an overlap in size between small cobbles and large gravel. Material that has been rejected as being too large for use as gravel can be purchased and used as cobbles.

The colour range is wide – wider in fact than that of setts, because the pinks, blues, and greens of metamorphic rocks can also be

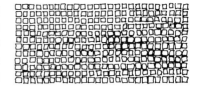
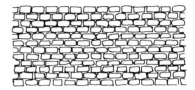
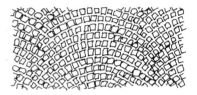

1.27 *Stone setts can be laid in a number of different bonds, depending to a degree on their shape. The fan pattern shown in the bottom example requires particular skills of the layer but is very effective when done well.*

included. But these more exotic colours cannot be specified with any confidence unless a source has been found, and the materials reserved, at an early stage of the project.

(ii) *Uses.* Cobbles are, probably, now used most frequently to deter pedestrians. When set with their longer axis vertical, and when there is a more comfortable surface to hand (or foot?), cobbles will keep most pedestrians away from areas where they might be unsafe or cause damage.

Cobbles can have a considerable decorative potential. Their use in both abstract and figurative patterns can be considered as a folk art well worth extending when budgets are high or when the project can be organized in a way that the labour of hand placing the stones is, in effect, free.

The small unit size, together with the decorative potential, makes cobbles particularly valuable when there are small and peculiarly shaped surfaces to be filled, but in which plants would be too vulnerable and a tar-bound or *in situ* concrete surface too prosaic. Their form also allows them to cope when the awkward shape also has awkward three-dimensional curves. Should this awkwardness have arisen, in part, because of the existence (or proposed existence) of a tree, then cobbles have no peer; they can be laid loose around the trunk (so allowing air and moisture to reach the roots) whilst being set as described below, where adjacent to other pavings.

(iii) *Construction.* From a technical point of view, laying cobbles is best considered as being an extension of that technique for exposing aggregates (described in section 1.13.8) which scatters the stones over the surface of *in situ* concrete and tamps them in. There are, however, two principle differences. First, the cobbles are not scattered but individually placed into position. It is at this time that patterns can be most easily incorporated. Secondly, if the cobbles are to stay in place they must be bedded to beyond half their depth. Only then will the matrix extend above their broadest part, leaving them a smaller hole from which to exit than they are able. If the concrete into which they are to be bedded already contains a high proportion of large aggregate, this would be very difficult. Moreover, that large aggregate might force the cobbles further apart than was the designer's intention. In such circumstances, the *in situ* slab should be allowed to harden and then be spread with a bed of 1:3 sand:cement, laid wet. The depth of this bed should be 50 mm (2"), or more if the cobbles are large. The objective being to finish with a bed depth that extends from above the cobbles' broadest part to about 25 mm (1") below them; approximately 75 mm (3") in most cases.

In areas that are never likely to take heavy loads, the mortar bed described above can be laid directly on a suitable consolidated base of hardcore or whatever. It might, then, be worth considering laying the cobbles in a dry-mix bed, and watering the area through a fine rose, when the whole has been completed. In that way there will be little risk of the stones becoming stained.

Figure 1.28 shows cobbles laid both on an *in situ* concrete slab, and on a hardcore base.

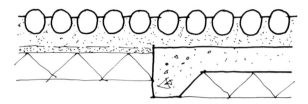

1.28 *Cobbles might be laid on consolidated hardcore or on an* in situ *concrete slab (as described in the text) depending on traffic loads. But note that traffic must be kept well back from the junction between these details or the bed to the cobbles will crack along that line.*

1.15.7 Reinforced ground cover

(i) Uses. Generally speaking ground cover plants can take only a very little trampling before showing adverse effects. Further, areas of planting could never take heavy vehicular loading even if those loads were only likely to be applied at very infrequent intervals. Yet there are often circumstances where the essential nature of a place demands a predominantly soft appearance despite the fact that pedestrian activity might be a little more than ground cover plants could cope with and/or provision might have to be made for occasional overflow car parking or for the possible passage of an emergency vehicle. In such circumstances the designer might well consider reinforcing the area in one of a number of ways which will resist wear or take infrequent heavy loads whilst allowing at least certain kinds of plants to grow and so soften appearances.

(ii) Types. As the above paragraph might suggest, there are two main classes of techniques which might be employed here. The first reinforces the planted surface so that it can take a slight increase in wear without a comparable increase in tear. The second reinforces the surface so that it can take occasional heavy loads. In both cases the plants with which they are most frequently used are grasses, but it does not follow that this must always be the case. If the designer has no specific horticultural understanding then expert advice would have to be sought. Unfortunately this book can go no further into this matter than to say that there are many plants other than grasses that can grow in very little soil and a whole range of effects could be achieved in reinforced soft areas.

(iii) Reinforcement against wear. A bare strip will appear in a lawn if too many people too frequently walk on the same line across it. With each step, leaves are torn and roots loosened. Too much of this

treatment and the plants die. But some grasses can resist more wear than can others. These, in the main, are those which form a matted sward as they spread vegetatively by striking roots from points on their stems. The stems become entangled in each other, and each is rooted to the ground in several places. Their natural resistance to wear is partially because such species tend to be quite coarse, but more importantly because of their mat-forming tendency. Such plants can have this natural quality further improved if the grip that they have on the ground can be enhanced. Two approaches are possible, and these might be used in combination.

First, several manufacturers make nets of plastic materials (which have some longevity) that can be pegged on to a surface prepared for sowing. This is then covered with a minimal depth of fine sifted topsoil and seeds of appropriate species sown. This net will itself become a structurally integral part of the mat of grasses. Unfortunately, there is no way of accurately predicting the degree of resistance to wear that this technique would give in any specific context. And, more unfortunately, if that proves to be insufficient then the deteriorated appearance with exposed net, is almost certainly less satisfactory than would be simple bare strip.

The second approach to reinforcing against wear is better in this respect. Many grasses will grow quite happily in a soil containing a high proportion of gravel or stone chippings. So, the area to be seeded might be prepared with 100 mm (4") or so of between 40/60 and 60/40 mix of topsoil and fine chippings graded between, say, 2–5 mm (0.08–0.2"). These chippings will also provide improved grip for roots (but not, it must be said, as well as would plastic nets). But if this does not prove to be good enough then, at worst, the result of the wear will be gravel path. This second technique could have a further possible advantage. The chippings would improve the porosity of otherwise poorly draining soils. That in itself is likely to improve resistance to wear. However, with soils that are already free draining, this could be a disadvantage in times of drought.

Figure 1.29 shows both of these techniques used in combination. Such a detail might be considered when anticipated wear is relatively high, when drainage might be improved with advantage, and when the fail-safe of a casual gravel path would not be aesthetically satisfactory. In the event that the detail failed, then the path would have to be entirely rebuilt in a more robust fashion.

1.29 *When grass is to be reinforced with a plastic net, the net must be securely pegged down into a topsoil/gravel mix, in order to provide a firm anchorage for the grass roots.*

Finally, it should be noted that the principal advantage of mesh-reinforced ground cover when compared with grass/concrete discussed below, is that there would be no colour or textural differences between reinforced and unreinforced parts of a single sward.

(iv)   Reinforcement to take occasional heavy loads. This branch of landscape technology has its roots in that approach to housing design which placed multi-storey blocks in a parkland setting. There was a concern to allow the parkland to run right up to – often right under – the flats. But there was also a concern that, in the event of fire, turntable ladders could be stationed right against the dwellings.

Hollow concrete blocks were laid on a hardcore base, and the voids in the concrete filled with soil and seeded. Whilst functionally successful, the visual concerns were never fully satisfied. Even though only a small proportion of the surface was concrete rather than grass, that grass as a whole always looked distinctly thin, probably because the compaction that had to be given to the base inhibited drainage and so the grass never really thrived.

Both precast and *in situ* concrete systems are currently available which exploit the principles of this technique. All are generally intended for lighter loads than described above. That, together with further research into the best forms for the concrete, gives much more satisfactory conditions for the plants, but, it must be said, that in no case could a sward run into and over an area reinforced in these ways, without there being a visual change in colour and texture. A proportion of the surface must be concrete and that must have visual consequences.

Figure 1.30 shows typical precast grass/concrete slabs. Other precast slabs, and *in situ* systems, will vary in detail but all follow the same principles. Manufacturer's catalogues should be consulted to establish specific recommendations.

1.30   *Grass/concrete slabs sit on a well draining sub-base. The voids contained within the slabs are filled with soil into which the grass is seeded.*

The subgrade must be firm, satisfactorily able to resist deformation (see section 2.6.5) and free of organic material. The formation (see also section 2.6.5) must be prepared at an appropriate level below planned finished levels. With lightly trafficked areas (overflow car parking for example) on a decent subgrade, a sub-base will not be needed. When a sub-base is required 100–150 mm (4–6″) will normally be sufficient. A 60/40 mix of granular material and topsoil might be used for the sub-base when loads are light, as this will promote healthy plant growth. A 25 mm (1") layer of sharp sand should be placed, screeded and lightly rolled on the sub-base or formation. Units are bedded into this sand with a tamping mallet. Cavities are filled with good quality, weed-free fine topsoil to within 25 mm (1") of the upper surface, and seeded with an appropriate mix. The area is then top-dressed with topsoil as above to within 12 mm (½") of the top of the units. Periodic light watering might be needed, in the absence of rain, to encourage germination. Unlike areas reinforced with plastic mesh, grass/concrete could be trafficked immediately.

1.16 TIMBER DECKING AND BLOCKS

Timber as a material, and ways and means of fixing and finishing timber, are covered more comprehensively later in section 2.6. The earlier parts of that section might usefully be read before that which follows here.

1.16.1 Uses

Timber, even when properly seasoned, has an unfortunate tendency to deteriorate rather rapidly. This would suggest that it is not an appropriate material for pavings. However, developments in timber preservation now make it reasonable to anticipate a useful life of up to forty years, even when in contact with the ground. Provided, as always, that the designer fully understands the limits of the material's potentials.

Timber shares, with clay bricks and stone, a near natural quality that induces a powerful psychological response. So, whether it is an overall strategy of integration or contrast that is being pursued, designers should only consider using timber in ways which allow that quality to be exploited. If its disadvantages can only be overcome by treating it in a way which destroys those qualities (thin tar coverings with scattered chippings to overcome slipperiness, for example) then some other approach using some other material should be adopted.

Timber will act like a sponge, soaking up water from the ground and air. This, in itself, can make it very slippery underfoot, and can encourage lichens and mosses which exacerbate the problem. The inclusion of herbicides and water repellants in the preservative treatment will help, but the only sound advice that can be given is that timber should not be used as a paving material in circumstances that are likely to be dank and airless, unless the paving is intended only to have a very short life, and so will not have time to become slippery. It must be accepted that, to avoid this problem, maintenance will both be regular and frequent, including dry scrubbing or that its applicability be restricted to contexts in which users are physically and mentally prepared for slipperiness (which, one suspects, is not often found in a 'space between buildings').

Whilst timber has been used for road construction, that application is not considered here. The notes which follow apply to footways only. Two fundamentally different technologies can be considered: decks and blocks.

1.16.2   Construction of decking

Timber decks are, essentially, bridges which are occasionally in physical contact with the ground. Indeed, one of the places where timber can be used most effectively as a paving, is where a bridge deck is allowed to run on beyond the element being bridged, to give unity through transition.

A minimum timber deck would have boards and bearers. 25 mm (1") should be considered as the absolute minimum thickness for boards in landscape work, and then only if the bearers are no more than 450 mm (18") apart. The width of the board is of less consequence, although it is worth noting that the small gaps (perhaps 3 to 6 mm) necessary between the boards, both to allow them to drain rapidly and to move with variations in the weather, will also provide grip. Narrower rather than wider might, then, be preferable. Boards should be fixed to the bearers either with nails (if the bearers cannot move) or screws (if stability relies on all parts of the deck acting as one).

The width of the bearers should be sufficient to allow the boards to be properly fixed. 38 mm (1.5") would be satisfactory if the boards are to run on over one bearer to the next, but 63 mm (2.5") would be needed where two boards have to butt above the bearer. The depth of the bearers is entirely dependent upon their span. A span:depth ratio of 25:1 should be adequate (e.g. a bearer spanning 3 000 mm (10'0") would need to have a depth of 125 mm (5"). The width of the bearer, to be safe, should be at least 30% of its depth.

The bearers could be placed directly on the ground (in which case their span would be nil). But that would only be satisfactory if they were purely temporary, and even then it is improbable that the

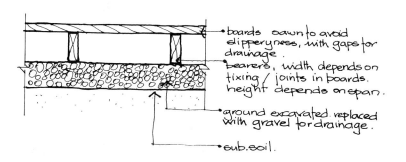

boards sawn to avoid slipperyness, with gaps for drainage.
bearers, width depends on fixing / joints in boards. height depends on span.
ground excavated. replaced with gravel for drainage.
sub.soil.

1.31 *A simple timber deck can be constructed with boards nailed to bearers which sit on a gravel bed to provide drainage. In some circumstances it might be advisable to treat the formation and the gravel with weed-killer.*

97

ground would be sufficiently firm and regular. It is more likely that, at a minimum, the area to be decked over has to be levelled and reduced by, say, 100 mm (4″). Stone chippings can then be consolidated over that area (giving a firm, level, and reasonably free-draining base) and the units of decking carried into place (Figure 1.31).

More frequently there will be a concern to have a minimum amount of timber in contact with the ground. It might then be better to set pairs of short posts into the ground to which the bearers can be fixed. The posts should protrude above the ground by a minimum of the depth of the bearer plus, say, 50 mm (2″) for clearance. The tops of each pair should be level with each other and at the level wanted for the top of the bearer. The posts would not have to be located at the ends of each bearer – indeed the mechanical strength of the bearer would be more fully exploited if it was to cantilever beyond each post by a dimension equivalent to 25–30% of its span between the posts. In this way the deck can be presented with literally no visible means of support, when that is desired. The loads which can be transferred between bearers and posts will be considerable. The joints here should be bolted rather than nailed or screwed. The depths to which posts have to go into the ground will vary with local conditions. It is most satisfactory when they can be driven in until firm, and then cut to level; for then both necessary strength and accuracy of level can be assured. If they have to be set in a dug or augered hole, they should probably go in by about twice the distance they protrude, or perhaps a little less if set in concrete. Figure 1.32 illustrates these points.

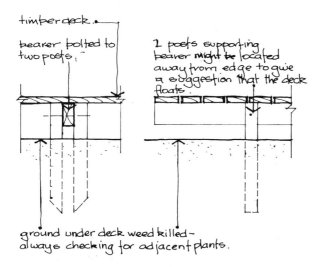

timber deck.

bearer bolted to two posts,

2 posts supporting bearer might be located away from edge to give a suggestion that the deck floats.

ground under deck weed killed – always checking for adjacent plants.

1.32 *A timber deck can be made to appear to that above the ground by supporting the bearers on timber posts set back from the edge of the deck.*

An alternative to posts (but only when the parts of the deck are screwed together to form integral units) would be to place the bearers directly on padstones. Where the ground is acceptably firm and level, these might be no more than a thick precast concrete flag which itself is set firm and level. If ground conditions demand it the flag (or small panel of bricks, or whatever) might be set on an *in situ* concrete base with a depth which takes it down to a reasonable formation. There would be no need for a hardcore or similar base below that. The units of decking are then simply lifted into place with a length of damp-proof material between the bearers and padstones. If each unit of deck is of sufficient size, they will stay in place through their own weight. If there is a worry that they might move, a short length of steel dowel (say 10 mm by 225 mm (⅜ × 11″) overall) might be set in the padstone to project vertically by, say, 75 mm (3″). The underside of the bearers can then be drilled to allow them to be dropped over those pegs (Figure 1.33).

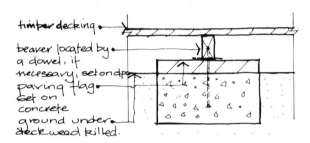

1.33 *As an alternative to the detail in Fig. 1.32 bearers might be located by a steel dowel set into a concrete padstone.*

There might be a concern (visual or technical) to have fewer posts. It would then be necessary to have beams between the posts, with the bearers (which now would be better called joists) running between the beams. There is no convenient rule of thumb for calculating the size of such a beam. A span:depth ratio of 20:1 is likely to be adequate if the spans of both joists and beam are, say, below 3.0 m (10′0″) and 15:1 if either is above that figure. If both spans are over that figure then advice should be sought from a structural engineer, as should be the case if beam depths of more than 300 mm (12″) were being considered.

There are a number of ways in which joists might be fixed to beams. Indeed, there are items of ironmongery (called, appropriately enough, joist hangers) available for this specific purpose. However, rarely is there the need to use this very strong detail, and even more rarely is it visually acceptable in landscape work. Two approaches which stay within timber technology are illustrated in Figure 1.34. Detail A is straightforward to construct and presents opportunities to variously profile the projecting ends of the joists. The depths of the notch in the underside of the joist must not exceed one-third of the overall depth of that member, otherwise there is a risk that it will shear. The principal disadvantage of this detail is that it almost certainly places the surface of the deck more than a convenient step-height above the ground. This would not be a problem if users were to be discouraged from walking on and off at will, but it would be if easy access was the aim. Detail B is a little less easy to construct, for the noggins must be very firmly fixed to the beam. The odd nail here and there will not do. A substantial screw immediately below the bearing of each joist would be a minimum. But modern wood adhesives have exceptional strength. If the element could be prefabricated, the noggins could simply be glued in the workshop and cramped in place until set. If undertaken on site, the adhesive should be seen as being in addition to screws, although because its job then is no more than to hold the noggin in place until the adhesive has set, the screw would not have to be so substantial.

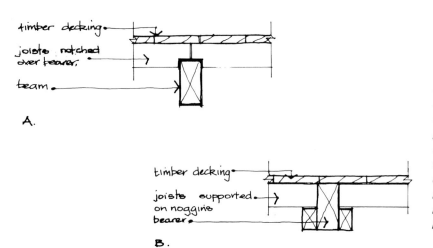

timber decking.
joists notched over bearer.
beam.

A.

timber decking.
joists supported. on noggins bearer.

B.

1.34 Where bearer centres have to be more than the deck boards could span, joists may have to be introduced. Simply nailing these to the top of the bearers might place the decking surface at an inconvenient height above neighbouring ground. Detail A shows how this might be reduced by notching the joists over the bearers, whilst detail B shows how this could be reduced yet further by fixing noggins on the sides of the bearers.

There is no intention of suggesting that these are the only details which might be considered; they are offered to stimulate rather than constrain.

Clearly, the posts supporting a timber deck can be taken up through the boarding to further support rails, fences, pergolas, etc. It should not prove difficult to bring together the information in this section and that on Fences (section 2.6) to produce an integrated design.

The timber used for the boards should be sawn not wrot. Smooth finishes are to be avoided to minimize slipperiness. The only exception to this would be when the boards were specially profiled so as to provide grip. The designer has a choice whether to use sawn or wrot timber for the joists and beams. If the timber will not be seen, it might as well be the cheaper sawn. If there was a risk of splinters catching people's skin or clothes, wrot timber should be used. That clearly is the case for any elements above deck level. All timber used for timber decking should be pressure impregnated to improve durability. This might not obviously affect the appearance of the timber, although those treatments which are most appropriate for this end-use will give a grey-green tinge. Alternatively, a pigmented treatment can be used, with a choice of any colour provided it is brown. All treated timber which is further worked on site must have those cuts or drill holes brush treated before assembly.

Timber which is not an acceptable shade after treatment, can be stained or painted on site. Stains are mostly in the brown and red-brown range. Some high-chrome greens, reds, blues and yellows are available. Whilst these all lack body they do allow the grain of the timber to texture the surface – the particular quality of stains. If the timber is to be painted, it must be wrot rather than sawn, but there is little point in painting the deck itself for the finish would soon wear away.

There is real value in treating the ground below a timber deck with herbicide to control weeds, but care should be taken towards the edges if plants are to be grown there.

1.16.3   Timber block pavings

Timber pavings can be made by laying blocks of wood directly into the ground. The timber can be sawn – in which case the blocks are frequently in the 100 × 100 – 150 × 150 mm (4×4 – 6×6") range – or whole sections of debarked tree trunks – with sizes from 150 to 400 mm (6–16") or more. Sawn timber is more vulnerable to decay as a very high proportion of its cells will have been cut through, but its smaller sections will allow more effective pressure impregnation. These advantages and disadvantages reverse with trunk sections. But, whichever variant is used, it must be remembered that the useful life can be quite short, for the walking surface will be fully exposed to the weather.

An area to be paved in this way should be excavated to a depth of 150 to 200 mm (6–8") depending upon the lengths of the blocks to be used, plus possibly a further 100 mm (4") if a lightly compacted base of stone chippings were to be included to improve drainage. The blocks are set with their long dimension vertical (so offering the end grain of the timber as the walking surface, which does give superior grip), packed in as tight as possible. Gritty sand should be thoroughly brushed over the whole of the surface – and this might have to be repeated at intervals throughout the life of the pavement. With trunk sections (where the gaps between each might be quite large) it would be best to ram the sand with a steel bar, in 100 mm (4") layers as those gaps are filled. Figure 1.35 illustrates this construction. Clearly it is critical that the blocks are set so as to achieve a level surface, but it is worth noting that the creation of informal flights of steps with timber blocks is relatively easy.

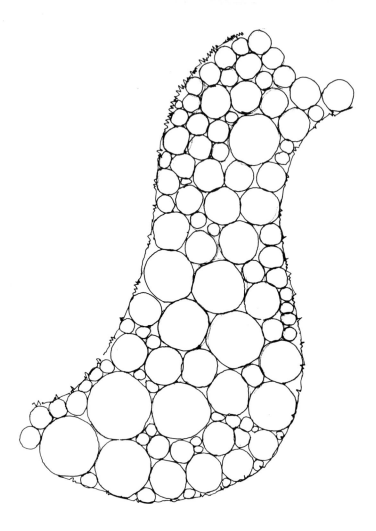

1.35 *Freely flowing areas of paving can be constructed from cylindrical sections of timber set vertically on a free-draining base. Interstices can be filled with gravel and/or topsoil.*

1.17 TRIM TO PAVED SURFACES

1.17.1 Edges and kerbs

Nothing establishes the character of a road more definitely than the treatment of its edges. Because both edges of a road are frequently a part of the same view, the kerbs can make a powerful contribution to the road's linear characteristics.

A tarmac or concrete road (materials with undoubted urban qualities) winding through the depths of the country, or travelling across open moorland, can be completely in character with its context if not divorced from those surroundings with upstand kerbs. The kerbs seem to provide a psychological channel through which town character can flow. Similarly, a gravel drive (a material which, in itself, has a rural informality) can be attuned to an urban context with crisp edgings of a kind that have no rural associations.

Upstand kerbs, moreover, can be dangerous where fast-moving traffic is expected. A motorist, who might otherwise have safely taken avoiding action, would almost certainly come to grief if a wheel touches an upstand. Further, pulling off the road to change a tyre or admire a view, could be impossible with an upstand kerb. Finally, on this particular point, draining surface water from a road with upstand kerbs requires a more complex and expensive technology than is the case when the edging is flush.

All of this has been said in order to illustrate that kerbs and edgings are (and should be considered as) an integral part of pavement design. They are conditioned by aesthetic, functional and technological criteria as much as are all other hard landscape elements. These must be understood in principle before the specifics of any individual case can be accurately judged and responded to.

1.17.2 Functions of a kerb

(i) *To prevent the lateral spread.* The way in which imposed loads are spread as they pass down through a structure, was described in section 1.6.6. Clearly if those loads were imposed very close to the edge of a pavement, they would try to burst out from the side of the construction (Figure 1.36). The effect of this on a granular pavement with an unrestrained edge would be for the structure to spread leading, sooner or later, to a breakdown of the structure as it becomes too thin to do its job. The edging (be it flush or upstand kerb) then, acts as a retaining wall, resisting those lateral forces, containing both them and the pavement structure.

(This is, in fact, not a problem with rigid structures which can contain the lateral stresses within themselves. But even then there is value in extending the base 200–300 mm (8–12″) beyond the *in situ* concrete slab to ensure that imposed loads are adequately spread over the formation (Figure 1.37).

(ii) *To assist surface water drainage.* An upstand kerb will be essential if storm water is to be taken to a surface water sewer via gullies. This will almost always be the case for public roads in towns. But in other circumstances in towns, and almost always in the country, drainage may be to linear drains, or to open or gravel-filled ditches (with or without land drains). In those cases a flush kerb is needed or, possibly, no kerb at all (Figure 1.38).

(iii) *To mark the boundary.* In an urban context where a footpath is to be provided adjacent to the length of a road, an upstand kerb will be reasonable because the change of level separates the two kinds of traffic. However, where both vehicles and pedestrians share a surface it might well be safer if that change of level were avoided, whilst retaining a purely visual statement by way of a warning to all that care is needed.

For example, in a pedestrian area to which a few vehicles were

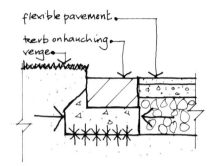

1.36 *The edging to a pavement has an important structural role to play. Not only must it spread vertical loads, as must the pavement as a whole, but it must also contain those forces which are trying to make the pavement spread.*

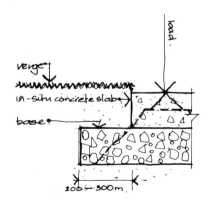

1.37 *An in situ concrete slab pavement, particularly if reinforced, does not have to have an edge restraint but, if not, then the base must be laid to project well beyond the edge of the slab in order to contain the cone of forces illustrated by Fig. 1.3.*

occasionally given access, a change of surface texture, or a demarcating strip set flush with both surfaces, would be sufficient to indicate what traffic was to be expected where. A change of level would be inconvenient and annoying, would suggest that some had more rights than others in particular places, and could well destroy the visual strategy through inappropriate tactics. In the same way, the occasional pedestrian and/or cyclist might be accommodated at the side of a road by some means which deterred, but did not physically prevent, motorists from running on to the edge. The single stone setts sometimes seen lining roads in Italy are an example of this. They are set at about 1 m (3'4") centres and are about 10 to 15 mm proud of the surface. They are intended to prevent motorists from driving casually off the road; the bumpiness of driving over them effectively prevents this (with the additional advantage of waking up drowsy drivers without killing them). This keeps pedestrians and cyclists safe without preventing picnickers and puncture menders from drawing off the road at slow speed. Moreover, the structure beyond the setts can be designed as appropriate for the lighter loads and slower speeds, so showing cost advantages as well. (All of which is well understood by motorway engineers who use cats-eyes instead of setts to mark the hard shoulder. But maybe the idea, in variety, has applications in more casual contexts?)

The visual definition given to the edge of the paving (particularly when traffic speeds are high) during adverse weather conditions, should also be considered. The use of reflectors and/or white lines can be invaluable when visibility is poor, but these can be incongruous in normal conditions. Better, perhaps, to use an edging, be it upstand or flush, that itself reflects headlights through the murk but with other characteristics which do not jar when bad weather is not dominating the situation. Similar thoughts can be given to the possible role of edgings when snow is lying. Knowing where the edge of the road is can be quite useful in such circumstances. Clearly an upstand kerb would only help in this respect when snowfall is light, but consideration of pavements and their edgings together with other associated elements (street lights, bollards, railings, planting) can and should also take in matters of this kind.

So, when designing a paved area, be it road, car park, playground, footpath, or whatever, it is important to decide first what character is to be created, preserved or extended and then to consider the functions of the edgings in that particular case. Only then will it be possible to decide what type of edge treatment, and its detail design, would be appropriate.

1.17.3 Types of kerb

(i) *Stone.* Relatively simple, but very tough and durable, edgings are made from granite (that is, building granite not necessarily geologists' granite). These are without doubt the most expensive of

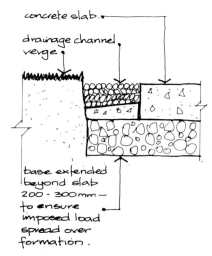

1.38 *The space above the projecting base (as illustrated by Fig. 1.37) can conveniently be used for a simple drainage detail. The channel must, of course, be taken at intervals to soakaways in the verge.*

the types considered here but have an all-pervading quality which more than justifies their use when budgets are sufficient. Their finish can vary depending upon the ways in which the stone is dressed. The smoothest finish is when fine picked. Medium texture is given when fair picked, single axed or nidged. The roughest texture results when rough punched. All are essentially rectilinear in section.

A wide range of sizes are banked (put into stock) at the quarries. Most permutations of 100, 125, 150 and 200 mm (4, 5, 6 and 8″) × 200, 225, 250 and 300 mm (8, 9, 10 and 12″) are available, and dressed for either longer and/or shorter sides to be exposed. Lengths vary between 375 and 750 mm (15–30″). Other sizes could be obtained to special order, within the limits of the material.

Radius kerbs are available in the same range of sections as above. They have an overall length of at least 525 mm (21″) and are curved so that arcs with the following radii can be formed: 1, 2, 2.5, 3, 4, 6, 9, and 12 m (3′4″, 6′8″, 8′4″, 10′0″, 13′4″, 20′0″, 30′0″ and 40′0″ in round figures). These radii are to the outside face of an upstand kerb and to the inside face of a flush kerb. This allows a flush kerb to be used as a channel block to an upstand kerb.

Quadrants are also available with a face radius of 450 mm (18″). These allow the kerbs to turn a soft rather than abrupt, right angle. Stone kerbs are laid in exactly the same way as are precast concrete kerbs (section 1.17.3 (ii)).

(ii) Precast concrete. As with flags, precast concrete is now used almost universally as an alternative to stone, simply because it is so much cheaper. Again, as with flags, the standard products give sound, if somewhat utilitarian, kerbs and edgings. They can be obtained pigmented and with exposed aggregates. Whilst this adds cost, they are still significantly cheaper than stone. The principal advantage of precast concrete (other than cost) over stone is, being moulded, a larger range of sections are available as standard. These, together with dimensions, are shown in Figure 1.39. All straight sections are manufactured in 915 mm (3′0″) lengths.

Radiused kerbs (with a half-battered section) and channel blocks are made, and, where these kerbs are asymmetrical about their vertical axis, both internal and external radiused sections are available. External kerbs are available to form arcs with radii of 0.91 and 1.83 m (3′0″ and 6′0″) and both external and internal kerbs to form arcs with radii of 3.05, 4.57, 6.10, 7.62, 9.14, 10.67 and 12.19 m (10, 15, 20, 25, 30, 35 and 40′0″). Figure 1.40 shows how these radii are to the face of the kerb. Quadrant blocks, profiled to match with half-battered sections, have radii of 305 and 455 mm (12 and 18″) and thicknesses of 125, 150 and 225 mm (5, 6 and 10″). Figure 1.41 illustrates quadrant blocks. Dropper kerbs are also manufactured which, in their 915 mm (3′0″) length make a transition from a 255 mm (10″) high half-battered kerb to a 150 or 175 mm (6 or 7″) high bullnosed rectangular section. Figure 1.42 shows a dropper kerb.

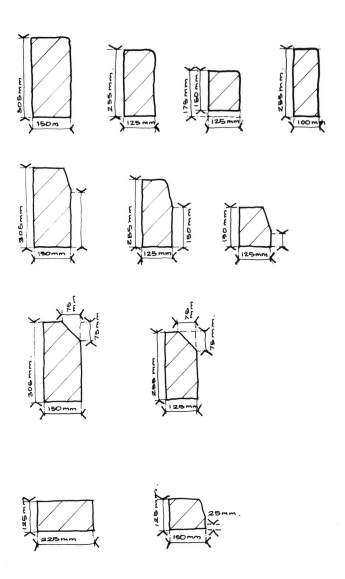

1.39 *Precast concrete kerbs are available in a range of shapes and sizes.*

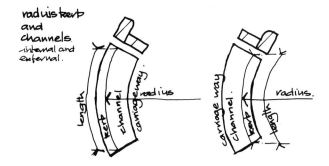

radius kerb
and
channels.
-internal and
external.

1.40 *Kerbs and channel blocks are available to allow the construction of bends with different radii.*

106

quadrants.

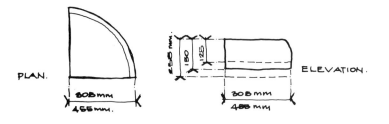

PLAN.

ELEVATION.

1.41 *Quadrant blocks allow a precast concrete kerb to be taken softly through a 90° bend.*

dropper kerbs. – these can be handed.

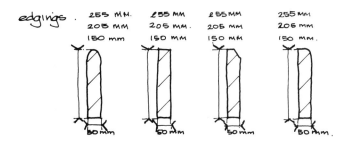

1.42 *Dropper kerbs are available to allow a transition between precast concrete kerbs of different sections.*

Edgings which are too insubstantial for normal use with roads, but with applications in association with pedestrian pavings, are also made. These are all 50 mm (2″) thick and 150, 205 and 255 mm high (6, 8 or 10″). They have half-round, chamfered, bullnosed or square tops as illustrated in Figure 1.43.

edgings .

255 MM.	255 MM	255 MM	255 MM
205 MM	205 MM.	205 MM	205 MM
150 mm	150 MM	150 MM	150 MM.

50mm   50 mm   50 mm   50 mm .

1.43 *Whilst generally of a consistent width of 50 mm (2″), precast concrete edgings for footpaths are available with different depths and profiles.*

Other sections and sizes are included in some manufacturers' product ranges. These are collectively too numerous to tabulate here. What can be said is that should a designer identify a difficulty with a particular detail that is of a kind that others are likely to have encountered before, then there is a very fair chance that there will be at least one manufacturer with a standard special which will neatly resolve that difficulty. As examples, kerbs are made which in their length change from one section to another (in a similar way as does a dropper kerb), and sections are available with mitred left or right-hand ends which allow abrupt (rather than curved) angles to be formed.

(iii)   *Setting stone and precast concrete kerbs and edgings.* The principles which underlie the details discussed here, also apply to brick kerbs and edgings – the differences are solely dimensional.

Kerbs, whether they are upstand or flush, and whether they have associated channel blocks or not, are always set before the construction of all parts of a flexible pavement above the sub-base. Indeed, they are an integral part of the construction of the pavement for they will contain the materials of the main body of the structure, whilst they are rolled. This means that the contractor must take care not to damage or dirty the edgings during the remaining construction, and the specification should make it clear that he has that responsibility.

Where the pavement has a sub-base it is convenient to extend the lower layers of that beyond the main body of the structure by a dimension that is equal to the width (or combined widths) of the edging(s) plus 300 mm (12″). This allows the edging to be set on an *in situ* concrete strip which extends some 150 mm (6″) behind the edging, which in turn allows a haunching behind (Figure 1.44). This haunching ensures that the edging is not shifted laterally by the outward and downward stresses that are inevitable during both construction and use. The minimum depth of the *in situ* concrete strip will be dependent upon anticipated loads. 100 mm (4″) would be sufficient in most cases but 150 mm (6″) might sometimes be used. Similarly, concrete with 14 N/mm² compressive strength would be sufficient in most cases but 21 N/mm² would be required when loads might be heavy.

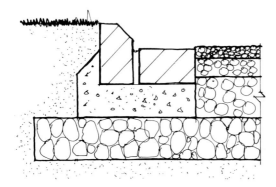

1.44 *Precast concrete kerbs and channel blocks are set on a minimum 100 mm (4″) concrete bed which acts as a foundation. This concrete is haunched behind the kerb to help it resist lateral loads. The concrete bed and haunch is often set on top of the sub-base when there is one.*

However, it is imperative that the overall depth of construction of the edging and its foundation is dimensionally related to the layers of construction in the main body of a flexible pavement. It is, at best, very difficult to consolidate a layer of granular material if it is required that it has a slight dinge or shoulder at its edges. Much better to reduce the depth of the concrete strip (but not below the minimum) and increase its strength, or vice versa, to allow each layer of construction to be consistently rolled. But, even more importantly, a designer should never consider simply adding unnecessary material

108

to the main body of the pavement as a means of adjusting these dimensions.

Often, indeed more often than not, the pavement does not need a sub-base, or that which is needed can be consolidated in a single layer. In those cases the concrete strip for the foundation for the edging is laid directly on the formation. In such circumstances there is only rarely any need to place a consolidated layer of hardcore or whatever below that foundation. The loads which come on to it will, at worst, be no greater than those on the main body of the pavement. The formation will have to be designed to be at a level which can resist those loads, and so it will also be adequate below the edgings. (The rare cases where a granular bed might be needed below the edging foundation are where there is a high water table and a likelihood of continued water percolation through the pavement surface. It could, then, happen that the edgings inhibit the drainage of that percolating surface water and so the granular bed is provided to give an outlet to neighbouring land drains. However, it must be added that in such circumstances it is probable that a sub-base has been introduced under the whole of the paved area for this specific purpose. So the situation returns then to that discussed in the paragraph immediately above.)

The concrete for these foundations is mixed with a minimum of water necessary to ensure it gains required strength (that is, rather stiff) and the edgings are set on to this immediately after it is placed. They are tapped to precise position and level with a wooden or rubber mallet, and the haunching concrete placed immediately after that. The edgings and their foundations then set as a piece, thus avoiding unnecessary labour and the risk that the haunch might not be integral with the foundation strip.

If there is any possibility of sulphate attack, sulphate-resisting cement should be used. It might then, if the foundation is set directly on the formation, also be worth providing an additional sacrificial 25 mm (1″) depth to the concrete strip.

Joints are normally butted with both concrete and stone edgings, but it might be with the latter that an expression of the individual stones is wanted, in which case 10 to 13 mm vertical joints might be left to be filled and pointed with a 1:3 cement:sand mortar.

With both flexible and rigid forms of pavement construction, expansion joints will be needed in the edgings and their foundations. With flexible pavements these should be at 18 to 20 m (60 to 65′) intervals. With rigid pavements they should coincide with the expansion joints in the pavement slab (see notes on expansion joints, section 1.13.7).

At least seven days should be allowed to elapse between setting the edges and laying the main body of the pavement; the edgings must be entirely firm if they are not to be displaced by those later operations.

The edgings to a rigid pavement might be set either before or after the slab is laid. When the edgings are laid before the slab it will, most likely, be because those edgings are to act as a permanent formwork for the slab. This, in effect, means that flush edgings are all that would be considered, for the top surface of the edging will be that which guides the tamping beam. In these cases, the edging will need exactly the same kind of foundation and haunch described above.

Conversely, an upstand kerb is most likely to be placed after the slab has been cast. The slab is given an additional 300 mm (12″) or so of width in order to accommodate the edging and a haunch. The kerb is set on a 13 mm (0.5″) 1:3 cement:sand bed. There can be a problem with this detail in that the haunch cannot be cast integral with the slab. If there is a probability that heavy vehicles will hit the kerb with some force a 75 mm (3″) length of 10 mm steel dowel can be pushed vertically into the slab, at 300 mm (12″) centres, at the time at which it is cast. Any tendency for the haunch to shear from the slab is resisted by those dowels (Figure 1.45). An alternative to this rather expensive detail is to use a rectangular section for the upstand kerb, set with its longer axis horizontal on a cement:sand bed. This block then acts as both kerb and haunch, with the additional width of mortar bed being sufficient to resist shear (Figure 1.46).

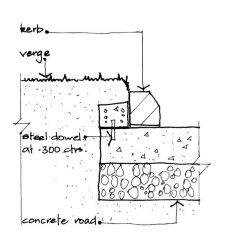

1.45 *With* in situ *concrete slab pavements, kerbs can be set directly on a mortar bed on the slab. Steel dowels are cast vertically into the edge of the slab to help the haunch behind the kerb resist shear.*

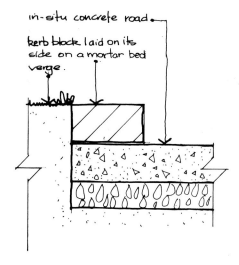

1.46 *As an alternative to Fig. 1.45 an upstand kerb to an* in situ *concrete road can be formed with a more substantial block.*

All of the principles discussed above apply equally to constructing precast concrete edgings to pedestrian pavements. The main difference in detail is that because these sections are relatively tall and thin, they need to be haunched on both sides to ensure they remain upright whilst the concrete sets (Figure 1.47).

No kerb should be designed purely in terms of its relationship to the pavement it is edging. It is not defining the edge of the world and so the way in which the ground surface is to be treated on its outside must also be considered. This is commonly another paved surface (footpaths beside a road, for example). A particular difficulty here can be ensuring that the road kerb has a sufficient haunch whilst, at the same time, ensuring that this is not so high behind the kerb as to occupy space needed for the footpath construction. The detail illustrated by Figure 1.48 is **not** satisfactory. The flags closest to the kerb are most likely to be run on by errant vehicles. Yet, with this detail, those flags would be given inconsistent degrees of support and so would certainly crack along a line immediately above the back of the haunch. Taking the haunch no higher than the footpath base will give those flags a consistent bed.

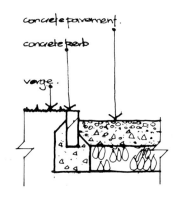

1.47 *Precast concrete edgings can be set on a concrete bed which is haunched up on both sides and used to contain an in situ concrete (footpath) slab while it sets.*

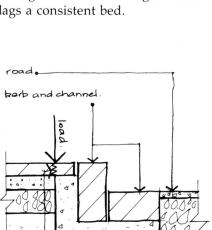

1.48 *This detail is **not** satisfactory. The haunch behind the kerb is also supporting the edge of the footpath flags. The flags are certain to crack should, say, a vehicle bump over the kerb.*

Equally commonly, areas to the outside of pavement edgings are intended for plants. (Trees in pavings are considered later.) If these plants are relatively large, there is rarely a problem, for their size requires that they are planted back from the kerb. But ground-cover plants (in particular) can find themselves with an insufficient depth of topsoil when growing immediately above the haunch. Again, this is the very place which is most likely to be trampled. It is not unknown for a muddy strip to develop between the planting and the kerb.

There can be further complications when the planted area is to be mown grass. Whilst small areas might be managed with a hover mower, it would be a real irritation (and unnecessary expense) to have to use such equipment for the edges of a lawn when the rest can only be economically cut with a larger, wheeled, machine. There would thus be value in including a narrow strip of light pavement between an upstand road kerb and an area of grass or other plants. A possible detail is illustrated by Figure 1.49.

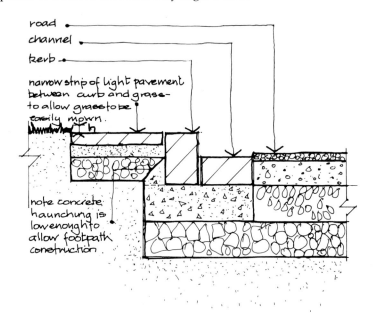

road

channel

kerb

narrow strip of light pavement between curb and grass - to allow grass to be easily mown.

note concrete haunching is low enough to allow footpath construction.

1.49 *Backing a road kerb with a course of flags can add visual emphasis to the junction between road and adjoining land. This will also aid moving when the adjacent ground is grassed, and avoid muddy patches should it be shrubbed.*

(iv) Brick kerbs. All brick manufacturers make a considerable number of standard special bricks (meaning that they are not the regular cuboids which are 99% of their production, but can be obtained ex-stock or after a short waiting period – special specials are made only to individual order). These are comprehensively illustrated in section 2.2. Some also have a value as pavement edgings, provided they are made from a material which is satisfactory as a paver (section 1.15.4). Further, some paver manufacturers include items made specifically for this purpose.

The more useful of the standard specials are bullnose, bullnose on flat, cants and header and stretcher plinths. Figure 1.50 shows these standard specials. These can be employed, in principle, in exactly the ways described for stone and precast concrete edgings above. There is but one further matter that the designer must consider. An individual brick has much less bulk and much less weight than a stone or concrete kerb. Designers must, then, be satisfied that a brick kerb or other edging detail will have sufficient substance for the job required of it. For example, Figure 1.51 shows a bullnose brick upstand kerb to an *in situ* concrete pavement. This would be entirely adequate if that pavement were for only light and slow-moving vehicles. However, if it were edging a road then it might be thought that it had insufficient vertical height, and that the width of the mortar bed was not enough to resist shear.

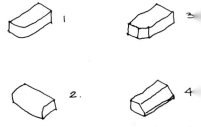

1.50 *Bricks are available in a range of standard specials. Four are illustrated here which might be useful for edgings of pavements: (1) bullnose; (2) bullnose on flat; (3) cant; (4) plinth.*

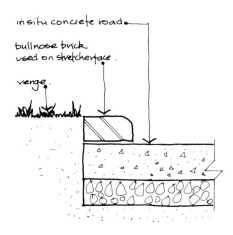

insitu concrete road.

bullnose brick
used on stretcherface.

verge.

1.51 *This detail shows a bullnosed brick used to form an upstand kerb to an* in situ *concrete road. Note that there is no haunching behind the bricks, so this detail would only be acceptable for light and slow-moving traffic.*

Pavers can be used on-end set in a double-haunched concrete strip (as described for path edgings, in Figure 1.47). For pedestrian-only surfaces these might have their width parallel with the path but if any vehicles whatsoever are likely to come into physical contact with them, their width should be perpendicular to the run of the road, in which case a single haunch might be sufficient.

Figure 1.52 shows special units made for use in association with interlocking block pavements. These are but samples and manufacturers' catalogues should be consulted for a comprehensive list. They could also be used to effect with pavers laid by more traditional methods – or indeed any other type of pavement – provided that details are designed within the principles discussed above.

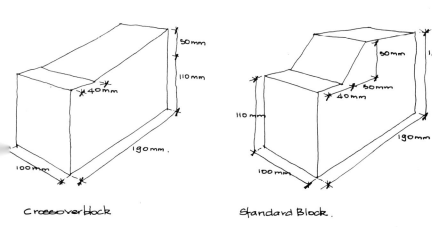

50mm

110 mm

40mm

190mm.

100mm

Crossover block

50mm 160mm

50mm

40mm

110 mm

190mm.

100 mm

Standard Block.

1.52 *Special units are available for use with interlocking block pavements. These are set on a concrete bed and haunch as in Fig. 1.44. The standard block illustrated here would give a continuous kerb while the crossover block would allow vehicles to pull off the interlocking block pavement on to say, a gravelled hardstand.*

113

(v) Timber edgings. Timber might not immediately suggest itself as a suitable material for pavement edging, even if its performance were improved through pressure impregnation. However, there are circumstances where the rate at which it deteriorates is of no significance and, in those cases, there is no need to go to the expense of special treatments.

The most common of such circumstances (which also usefully illustrates the point being made) is the edging to a cold-rolled asphalt (or other light, tar-bound) footpath. At the time at which it is constructed, it is critical that the edges be restrained so that layers can be consolidated. Further, it is most probable that neighbouring surfaces will not have been formed at the time the path is laid. Indeed, those surfaces may not be there at all; if they are to be planted and the topsoil has been stripped off (as it should have been) to protect it during the building period. However, once the path has been constructed and the adjacent grass (or whatever) has had time to form a dense sward, the edging will no longer be needed. Given that an informal junction is what is required between path and grass, then there is no need to spend more on the path edging than is necessary to give (say) a five year life. Timber, in these cases, is ideal. Fixed upright with 450 mm (18″) softwood pegs at 1 m (3′4″) intervals, as shown in Figure 1.53, it will contain both base and finishing course during path construction, and ensure there is adequate depth of topsoil for plants. True, it will ultimately rot away but not at so fast a pace that the voids it creates cannot be filled by other natural processes.

The width of the board should be that of the path construction, plus (possibly) a small upstand. The thickness should reflect the loads which will come on it as the pavement is consolidated, but 32 mm (1¼″) should be satisfactory in most cases, and as little as 25 or even 19 mm (1″ or ¾″) for very light paths.

A particular advantage of this technique is that free-form curves can be made with relative ease, and the thinner the boards the easier this will be. Additional pegs may have to be set if the nature of the curve means the board would spring prior to consolidating the footpath, and shallow vertical saw cuts may have to be made in the board where the radius of the curve is tighter than the board thickness can manage. Further, it is this advantage which might suggest the use of timber boards to shutter the edges of a freely curving *in situ* concrete path. Clearly, no attempt can be made to reuse this shuttering, but then the probability is that there are no other identical curves.

(vi) Other approaches to kerbs and edgings. No attempt could be made to give this section comprehensivity, nor would it be possible to give it much structure. That being so, no such attempts have been made. It is intended to do no more than illustrate some less commonly encountered edging details which, by their nature, have only limited

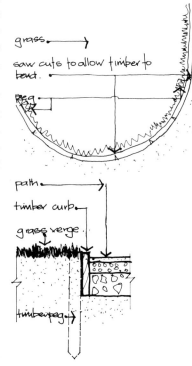

1.53 *Timber planks can be used to edge light bound-macadam pavements. Kerbs can be formed that would be difficult, if not impossible, with other materials. By the time that the timber edging has rotted away the macadam and the ground behind the edging will be well consolidated.*

applications but which can do a great deal to establish individuality and character – all with a view to encouraging designers to use their skills to the full when considering edgings and, perhaps, less frequently simply to adopt the most convenient conventional approach.

It is not unknown for designers working on nice flat drawing boards to lose sight of the fact that the place they are designing for is not, itself, flat. This is particularly the case where existing falls are not great and so are having no radical effects on the building design. But even very slight falls can be used to advantage at pavement edgings, the very places where level differences are likely to collect. Figure 1.54, for example, shows a stepped kerb detail which gives the pedestrian a very satisfactory feeling of being both physically and psychologically separated from, and superior to, the motor traffic.

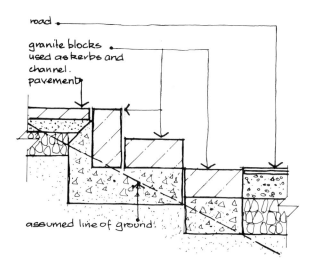

1.54 *Kerbs and edgings do not have to be purely utilitarian. In this example a change of level is exploited to give an edging with considerable strength of character.*

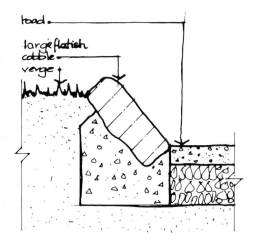

1.55 *Provided that the functions of kerbs are understood, designers can work successfully with materials which might not commonly be employed for that purpose.*

115

Figure 1.55 is informed by a memory of a detail seen in Denmark many years ago. (Being so long ago, the memory may not be entirely accurate but will serve to illustrate the point.) The coastline along which a very lightly trafficked road was running was littered with large flattish cobbles deposited, one suspects, by retreating ice. The use of these as a simple edging effectively related the road with the landscape of which it was a part. This approach to design must have applications in many contexts, not solely rural and not solely where natural processes provide inspiration and/or materials.

It is possible to construct a bound flexible pavement without any edgings provided that there is sufficient space at the sides to allow each layer of construction to run out beyond the extent of that above, and provided that heavy vehicles are not likely to regularly want to run very close to those unsupported edges (Figure 1.56). This has advantages that are not only economical but also visual. An irregular line develops between road and verge promoting a casual relationship between hard and soft elements. But such informal characteristics are, technically, more difficult to achieve when vehicles must be kept back from the edge for their own protection, and/or for that of the pavement. Figure 1.57 shows a way in which such a combination of difficulties has been resolved in a country park. The pegs shown in this detail ensure that the logs are not accidentally rolled out of position if touched by a wheel. They could have been avoided had half logs been used. But those would have had to have been much heavier if their height was to be effective, would have had a substantial surface in contact with the ground (and so would have deteriorated more rapidly), and would have been visually much less

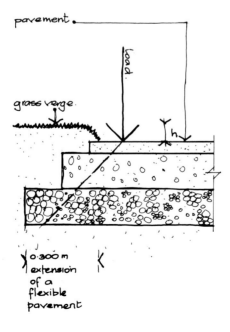

1.56 *Rural roads often look best with no kerb edging. The various layers of a bound macadam pavement must each extend beyond that immediately above by a dimension that is at least equal to their own thickness. In this way the spreading forces from an imposed load will be continued by the structure as a whole. The grass verge should be set 50 mm (2") or so above the surface (h), so that the edge of the wearing course is hidden.*

116

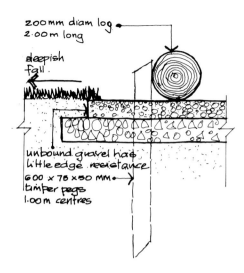

200mm diam log
2·00m long

deepish
fall.

unbound gravel has
little edge resistance
600 x 78 x 50 MM
timber pegs
1·00m centres

1.57 *Kerbs should be in character with the general surroundings. This detail was seen in a country park.*

effective. Again, it must be possible to re-interpret this notion for a more intensely used urban situation.

Finally, many human activities have, over the centuries and by their essential nature, become associated with particular materials used in particular ways. These associations can be exploited both when designing for such activities and when wishing to borrow those characteristics for application elsewhere. Figure 1.58 shows how bullnosed engineering bricks might be used as an edging with a visual strength given as much by association as by that material's actual qualities.

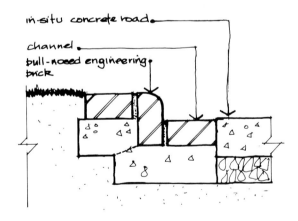

in-situ concrete road.

channel.
bull-nosed engineering
brick

1.58 *The edging to this in situ concrete pavement has been designed in such a way that the engineering brick channel on its concrete bed both contains the concrete pavement whilst it sets and gives a true surface to guide the tamping beam, whilst the bed to the mowing strip also acts as a haunch behind the upstand bullnosed kerb.*

1.17.4 Surface drainage

It would be beyond the scope of this book to consider the design of

117

surface water drainage systems as a whole. This section will, then, concentrate its attention on those parts of such systems which show at ground level.

Only the lightest, unconsolidated gravel pavements allow rain-water to be simply percolated through them to drain away into the ground. All other pavement structures are, or very rapidly become, impervious and so must be designed with surface water drainage in mind. The object of the exercise is to ensure that rain-water will run off the paved surface as quickly as possible without unduly compromising the other necessary qualities of that surface, and conducting that water to places where its presence is not an inconvenience.

(i) *Falls to paved surfaces.* If rain-water is to run off a paved surface, that surface cannot be flat and level but must be laid to falls. Roads and paths may be cambered (have a surface profile that is a low arc, giving falls to both sides), have cross-falls (a straight fall to one side or the other), or in some rather special circumstances have straight falls from the sides to a central gutter. But areas of paving often have a width such that it is not possible to take all surface water to their edges without that water building up into an unacceptable volume. These, then, must have a system of falls taking water to channels and gullies within their area.

Different types of pavement surfacings have different minimum falls if rain-water is to run off at a rate sufficient to ensure that depths are never such as to be dangerous for vehicles or inconvenient for pedestrians.

An *in situ* concrete pavement can only be built with straight cross-falls, for the beam with which it is tamped is itself straight (see section 1.13). Where the ridges which result from tamping are running with the fall (by far the most likely circumstance) then minimum lateral falls of 1:60 are recommended in association with longitudinal falls of 1:100 or, at an absolute minimum, 1:150. If surface texture is more likely to impede rather than assist drainage, then minimum lateral falls of 1:50 or even 1:40 should be considered.

A bound macadam road could be cambered or with straight cross-falls. With a cambered surface the minimum recommended lateral fall is 1:48, and that for a cross-fall (where the length of fall will be the whole, not a half of the width of the road) is 1:40; both associated with minimum 1:200 longitudinal falls. Where an area of bound macadam has a system of falls, the rates of fall need to be interpolated within these parameters, that is, with a length of fall of 1.5 m or thereabouts (say 8'0") a 1:50 fall would be acceptable, but with a length of fall of 5.0 m (say 16'0") a fall of 1:40 should be considered the minimum.

With interlocking concrete block pavements, a minimum lateral cross-fall of 1:40 with longitudinal falls of 1:100 are recommended wherever practicable. These are also the figures recommended for flexible brick paving and for rigid brick surfaces for roads. But minimum lateral falls of 1:60 are acceptable for pedestrian-only pavements where rapid drainage is less critical.

An unbound macadam surface where significant percolation is to be avoided would need minimum lateral falls of 1:30. Such surfaces are likely to discharge directly into a French drain and so minimum longitudinal falls are not quoted.

Quoting minima for the falls on precast concrete and stone flags, and for setts and other small unit pavings, is very difficult. Not only is there great variety between these structures but there is also great variety within them. Designers must reach their own conclusions in the light of the specifics of each case, remembering that the critical variables are the smoothness of the surface (the smoother it is the faster the water will flow), the length of fall (the greater this is the more the water that will collect towards its lower parts), and the function of the surface (a depth of water on a road could be dangerous whilst on a footpath merely inconvenient). So it is that a wide setted road with cross-falls might need to be designed to minimum 1:40 falls whilst a narrow footpath built of large untextured precast flags could have falls as low as 1:120.

Stating maxima for falls is also rather difficult. These are conditioned by the functions of the surfaces not their construction. Longitudinal falls of more than 1:10 will present motor vehicles with difficulties, particularly when icy. But we all know of hills of 1:4, if not steeper over short stretches, that are (rather slowly) negotiated by heavy lorries. The danger is rather greater when the vehicle is descending a steep road. The stress on the brakes and the possibility of loss of adhesion are both considerable. It is for this reason that some planning authorities quote 1:12 as the maximum longitudinal fall for even quite short ramps and drives which arrive directly at their junction with a public road.

Maximum longitudinal falls for footpaths are more a matter of psychology than physics. A slope which might actually be perfectly safe, may well not appear to be so. It is worth noting that a flight of steps will have a rise:go relationship of about 1:3 whilst a perron (a very convenient device discussed in section 3.3.3) accommodates a slope of about 1:7 with the sloping treads themselves having a fall of, say, 1:15. So, it might be concluded that if longitudinal falls on a footpath were much over 1:10, the designer should start to consider the use of steps rather than ramps, given, it must be stressed, that alternatives can be provided for people in wheelchairs or with prams.

It is beyond the scope of this book to discuss maximum lateral falls for roads where traffic speeds might be high. This is a matter of some complexity (particularly at bends) and one where expert advice would be needed. With the kinds of short access roads that are within the scope of this work it would be safest to consider minima as maxima. That is to say, if the specifics of a particular design are leading towards conclusions that give lateral falls to a road that are greater than 1:40, particularly if they are cross-falling from the inside to the outside of a bend, then designers should investigate the possibilities of taking up some of that fall with an adjacent retaining wall, and/or seeking the assistance of a civil engineer.

Maximum lateral falls on footpaths are, again, largely a matter of psychology. Cross-falls of less than 1:50 seem, generally, to go unnoticed. However, once those falls become steeper than around 1:30 they are definitely perceptible and could well worry the old or infirm; and the problem would be compounded if the path was surfaced with a material that looked a little slippery even if it was not.

It is not possible to offer specific advice about maximum falls in areas of paving, be they for vehicles or pedestrians. So much is conditioned by circumstances. Falls which might well be acceptable around an infantry barracks are likely to be totally unacceptable associated with a hospital. But one general point can be made, and this seems frequently to be overlooked in many matters of design, not only gradients. It must be remembered that surfaces designed as car parks also have to be used by pedestrians getting to and from their cars. Falls should always be designed to satisfy the most constraining of the applicable criteria.

Finally, designers should take care to avoid the trap of considering the falls appropriate for different but adjacent areas of paving, individually and sequentially. Whilst designers might deliberately introduce a step between two areas of paving, it is totally unacceptable for that to happen inadvertently. The levels and falls at the edges of abutting pavements must be identical where a step is not deliberately intended and, if a step is intended, it must be in the order of 100 mm (4″) if users are to appreciate that it is there and not trip up. Small surface misalignments, particularly where height differences increase along a line, are very dangerous.

(ii) Channels. Channels collect the rain water draining from a paved surface and direct it to a gully. They might be used in two kinds of places: adjacent to an upstand kerb where the paved surface falls towards that kerb, and at the line of junction between two areas of paving falling towards each other.

To this must be added that with bound macadam pavements a channel block may be needed between the pavement and an upstand kerb, should it be necessary to use a heavy roller that could not be steered right up to the kerb. Also, channels are often and effectively incorporated into a paving design when, strictly speaking, the volume of run-off does not justify their use, but when some over-emphasis of a technical solution assists with visual strategies or tactics (see section 1.1).

Channels can be dished or flat, and can be formed of any paving material which will not be washed away by running water. Special dished slabs are made in stone, brick and precast concrete. These look (and are) workmanlike and efficient; but stone dished channels are very expensive and so are not often used. Dished channels can be formed *in situ* using bricks, concrete blocks, setts or cobbles. The majority of channels used (particularly in association with roads) are flat precast concrete. Flat channels can also be formed *in situ* with all of the other materials already mentioned in this paragraph.

Flat precast concrete channel blocks, their forms and sizes, and the ways in which they might be used, have been discussed with precast concrete kerbs in section 1.17.3. Similarly flat stone channels were considered together with stone kerbs in the same section.

Setts have been widely used to form dished channels in the past. Figure 1.59 illustrates a typical detail in association with an upstand stone kerb, and Figure 1.60 shows one way in which dished setts might be related with an area of flags. The use of setts for channels is a reasonably economical way of introducing a quality product into a paving design, but they can be very annoying for cyclists trying to keep close to a road kerb, away from motor traffic.

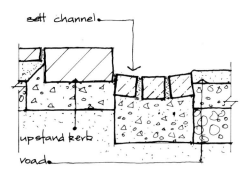

1.59 *Stone setts, being relatively small units, can be formed to use a dished channel. Their qualities as a material will significantly enhance otherwise purely utilitarian areas.*

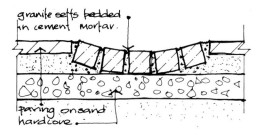

1.60 *This detail is* **not** *satisfactory even if vehicular loads are not anticipated; the setts should be given a minimum 25 mm (1″) mortar bed. Either the bed below the flagging must be increased or the hardcore base laid locally at a lower level.*

Cobbles are less than ideal as a material for forming channels. Their irregularity inevitably slows the rate of flow. However, they can have a use not so much as a channel but as a continuous gully grating over a simple linear drain. Figure 1.61 illustrates a detail which might be considered in appropriate circumstances.

Channels formed of bricks, whether they are dished or flat, are best built as a narrow strip of rigid paving (that is, bedded on mortar on an *in situ* concrete strip) if there is any likelihood that they will be traversed by motor vehicles. This is so whatever the neighbouring paving. The only exception to this advice would be if the channel were no more than a running bond border to an interlocking pavement. In such a case the channel is more notional than actual and so can be laid with the pavement surfacing as a whole. However, in pedestrian-only areas the blocks can be used to form channels laid on a full bed of mortar directly on to the pavement base. Figure 1.62 illustrates this possibility. All that might usefully be added to this is that where rectangular units are used, and where drainage efficiency is to be optimized, the blocks are best laid in a running bond directed towards the gully gratings.

Some precast concrete flag manufacturers include dished flags within their product range. These are generally a little thicker than the standard flags with which they are used, so as to accommodate the depth of dish. They can be laid in exactly the same way as has been adopted for the neighbouring flags, given that they are on a bed which is sufficient to accept the slight additional depth of the channel flag (Figure 1.63). Alternatively, channels might be formed with small unit flags used in an essentially similar way to that discussed and illustrated for bricks.

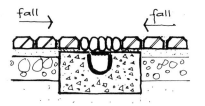

1.61 *The appearance of this linear drain has been improved by covering it with cobbles laid loose. The cobbles are contained by paving bricks bedded on mortar on the concrete surround to the drain. The remainder of the pavement is pavers vibrated on a sand bed.*

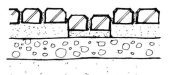

1.62 *This dished channel has been formed by setting pavers on a mortar bed on hardcore, whilst the rest of the paving is identical pavers vibrated on a sand bed. Note that this detail might not be considered satisfactory if the pavers did not have chamfered edges (see B, Fig. 1.63).*

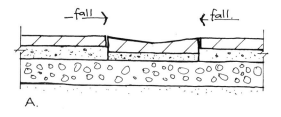

A.

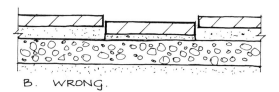

B. WRONG.

1.63 *Detail A shows a dished channel flag set on mortar on a hardcore base, with the adjacent paving being flags set on a cement/sand dry mix. Detail B is* **not** *satisfactory; very small upstands are difficult to see and, in consequence, very easy to trip over.*

(iii) Gullies and gully gratings. If surface water is to be disposed of through an underground piped system, then it will have to be taken

off the paving and into that system through a grated gully. The grating inhibits the passage of any large objects whilst the gully traps smaller material which might otherwise silt up the pipes. Whilst the design of the underground parts of a drainage system are beyond the scope of this book, Figure 1.64 shows diagrammatically the relationship between grating, gully and pipe, so as to set these paragraphs in their broader context.

Gully gratings are made in an almost bewildering range of shapes and sizes. Square, rectangular, circular, triangular, double triangular and multiple triangular, are all available, most with alternative arrangements of bars, and in cast or ductile iron, or steel. Clearly all these cannot (and need not) be discussed here. What the designer must understand is that they vary in strength. There are three main grades. Grade A includes heavy duty gratings and frames for use in roads with high speed and heavy vehicles. Grade B gratings have applications in minor residential and access roads where heavy vehicles are rare and slow moving. Grade C products can be used only in areas free of any vehicular traffic. Different manufacturers add to these classifications in different ways. One, for example, includes grade A+ which will take very heavy loads such as aircraft; grade A2 with a value in car-parks; grade A, class 2 for use in pedestrian areas to which heavy delivery vehicles have limited access, and grade C+ for use where light (but never heavy) vehicles might be allowed. The designer must decide what grade, shape and size is wanted and then make reference to manufacturers' catalogues for precise specifications.

Gully gratings are available which co-ordinate dimensionally with standard precast concrete and stone channel blocks. Clearly there is both technical and visual advantage to be gained from such co-ordination. Indeed, when designing a channel to be built *in situ*, it is well worth developing the detail for the channel with a gully grating catalogue to hand.

Gratings of the same shape and size are available with different bar patterns, flat or dished, hinged and/or lift-off, and with different seating arrangements. Bar pattern selection is likely to be influenced by drainage volumes and directions. Figure 1.65, for example, shows a pattern developed specifically for use with steep longitudinal falls where flow rates are likely to be high. Figure 1.66 shows a pattern with a value within an area of paving where water flow is from a number of different directions. Clearly, flat and dished gratings are intended for use with flat or dished channels.

Because gully pots trap fine material which might silt up the underground pipes they have to be cleaned out at intervals to ensure that they themselves do not silt up. With larger gullies it may be necessary to fit a lift-off grating so as to maximize access space. With smaller gullies it may be sufficient for the grating simply to be hinged. In certain circumstances using gratings (hinged or lift-off) with an anti-vandal device may be advisable.

Different seating arrangements are intended for use with different

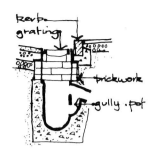

1.64 *The frame to a gully grating is bedded on mortar on brickwork. The number of courses of brickwork will depend on the relative levels of the drain and the grating. The brickwork is built off the concrete surround to the gully pot.*

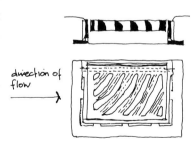

1.65 *This storm gully grating is recommended for use with longitudinal gradients of 2% (1 in 50) or greater. The curved pattern of the bars helps to slow the water flow which otherwise might simply rush straight over the grating.*

weights and densities of traffic. With light, low speed traffic, special precautions are unnecessary. But with higher loads, speeds and densities, gratings that will not rock are an advantage to drivers, vehicles and gratings. Figure 1.67 shows such a grating.

All the gratings discussed so far in this section are intended to be built flush with (or, rather, fractionally lower than) the pavement surface. However, there is an alternative approach intended for use with upstand kerbs. Figure 1.68 illustrates variations of these. Outlets have a face profile which matches with square, half-battered or splayed standard precast concrete upstand kerbs. The water is taken off horizontally through the face of the kerb. The simplest (known as kerb weirs) act like the scuppers on a ship – the water simply drains through the open back into the ground (or gravel filled ditch) behind. Others have a spigot allowing the water to be piped away from the pavement, again to a ditch, soakaway or similar (or, possibly, a remote gully pot). The more complex are intended for use with a gully pot located behind the kerb rather than under the pavement. The solid cover, which can be relatively lightweight, gives access to the gully for cleaning. All or any of these might be used in association with an anti-flood deflector built into the channel. This ensures that, with steeper longitudinal falls, the majority of the water does not flow straight past the weir.

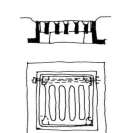

1.66 *Square gully gratings such as this are particularly useful in central positions within pavements where no particular direction is to be emphasized.*

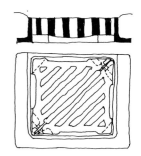

1.67 *This heavy non-rocking gully grating and frame would most likely be employed in a heavily trafficked road.*

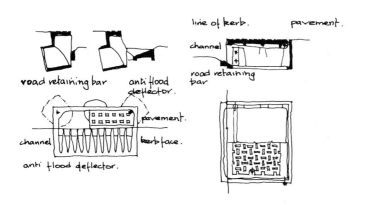

line of kerb.   pavement.

channel

road retaining bar

road retaining bar   anti flood deflector.

channel   pavement.

kerb face.

anti flood deflector.

1.68 *To the left is illustrated a kerb weir with anti-flood deflector. The detail to the right is similar but also allows access to the gully beneath.*

(iv) *Linear drains.* Gravel-filled ditches, with or without land drains, are the simplest possible linear drains. However, they are of limited value within, rather than at the edge of, a pavement, where they would be very inconvenient for pedestrians and could well adversely affect the pavement's foundations.

A pavement built with a system of falls leading to individual gullies without channels will undulate on two sectional axes. And a pavement designed to avoid these complex undulations by the use of channels will have its surface interrupted by the dishing or depression of those channels. In circumstances where neither of these conditions are acceptable, linear drains of some kind would be the only remaining option.

Linear drains can be grated channels. The grating itself might be level (with the adjacent paving cross-falling to it from either or both sides) in which case the channel beneath must be laid to falls. Alternatively, the grating can be laid to fall with the pavement surface, in which case the channel beneath can have a constant invert below the grating.

Linear drains can be no more than a continuous slot in the pavement surface that allows for the water to gain access to the channel beneath. Again the channel might be laid to falls or have a constant depth below the surface, dependent upon whether or not the surface is itself falling on that line.

Both these forms of linear drain can be built *in situ* or use one of a number of different proprietary systems. Whatever the case, the water in the channel must be conducted through a gully to a surface water sewer, or to a soakaway. With the proprietary systems, all the necessary special units which allow such connections are available. When linear drains are built *in situ* the channels may have to be run into an inspection chamber from which the water is piped.

Figure 1.69 shows a slotted linear drain built *in situ* within an area of rigid brick paving. This detail would allow the pavement on the line of the slot to be level whilst the channel below is laid to falls. Figure 1.70 shows a proprietary grated linear drain, also in an area of rigid brick paving. These proprietary units come in standard lengths of 500, 750 and 1000 mm (20, 30 and 40″). Each incorporates a 1:166 fall within their length and inverts (the vertical dimension between channel and surface) vary between 125 and 300 mm (5–12″). A number of units within this range have knock-out panels allowing spigots to be fixed, through which the water can be piped to gully or soakaway. Such a system, together with associated silt traps, petrol interceptors, flushing arrangements, etc., has all the flexibility needed to drain any kind of pavement. Slotted linear drains look very neat and tidy but can be rather difficult to clean. Conversely, grated channels have greater visual impact but are much easier to maintain.

(v) Inspection chamber covers. Most paved surfaces have inspection chambers below them which allow maintenance access to surface water drains, to foul sewers and to other services. The covers to these also impose both technical and visual constraints on pavement design.

Like gully gratings, the covers and frames are made in a wide range of shapes, sizes and strengths. They are available in both cast and ductile iron, and steel. Whatever other decisions might be taken, covers must be selected that are appropriate for the loads and speeds of the traffic they will have to withstand.

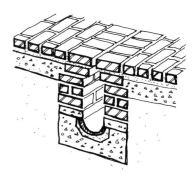

1.69 *This detail shows a linear drain built* in situ *within an area of rigid brick pavings. This type of drain would be almost invisible on the surface but does present maintenance difficulties. Adequate provision must be made for the removal of silt. The inclusion of access covers for this purpose could defeat the visual objective if these covers were not located with due care.*

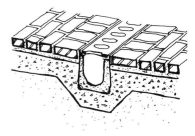

1.70 *Proprietary linear drains do not have the maintenance difficulties associated with the drain in Fig. 1.69 because their surface is a continuous and removable grating. However, not all of these gratings (there are several different patterns) are entirely easy on the eye.*

125

Drains and sewers have to run in straight lines and at constant falls between gullies and inspection chambers. Moreover, there are maximum distances at which they can be set. The pavement designer does not, then, always have a great deal of choice as to where those chambers and their covers will be. Whilst this is of rather less consequence when the pavement is bound macadam or *in situ* concrete, which read as a continuous surface, much more care must be taken when the surfacing is brick, flag or other small unit material, particularly if it incorporates a pattern. If left to chance, Murphy's law guarantees that the cover will turn up at that place which does most visual damage to bonds and patterns. In such circumstances, the use of recessed covers might be considered. These are available with a range of depths of recess, allowing the cover to be filled to match adjacent paving materials, as illustrated by Figure 1.71.

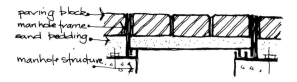

1.71 *Recessed covers are available for manholes and inspection chambers. These will accommodate the materials used for the adjacent pavings – but note that there must be dimensional co-ordination between the covers and the paving units.*

Even when set in a continuous paving surfacing, inspection chamber covers often seem to do more visual damage than should be necessary. Rarely is it critical to the drain that the cover be orientated in any particular direction. (The chamber below, more often than not, is circular.) It should, then, be possible for the cover to be set so that it is at least parallel to neighbouring buildings, kerbs, parking-bay indicators, or whatever. Such a simple matter would take little effort to achieve and add nothing to the cost of the job, but would make a significant contribution to general tidiness. In this respect, circular covers have a particular value, for they have no particular orientation.

The design and construction of the chambers themselves is beyond the scope of this book, but it should be noted that the cover and its frame are set on the brickwork or *in situ* concrete surround to the chamber. The importance of this, for the pavement surface, is that the pavings may well have better support immediately adjacent to the cover than they have elsewhere. This is of little consequence with macadamed surfaces which can flex over this discontinuity. But an *in situ* concrete slab might well need light mesh reinforcement (if it is not already reinforced) to give it the strength necessary to bridge between chamber foundation and the pavement base. And brick, flag, or other small unit paving must be bonded so that no individual units attempt that bridge; if they do they will surely break under any loads greater than pedestrian.

1.17.5 Trees in paved surfaces

It is extremely difficult to know in just what circumstances a tree will thrive in a hard area, since it is seldom possible to compare the soil conditions under different pavements. Soil which might be expected to be bone dry may be kept moist by an underground spring; soil that was once moist throughout the year may have lost its catchment area (which is not necessarily only the surface immediately above it) due to the construction of a car-park or new road. When walking about a town one can see healthy, flourishing trees which have, apparently, no special provision made for them, and poor feeble specimens upon which every care has been lavished. Almost every tree in Paris has its own circular iron grating whilst in London these are relatively rare. Yet splendid trees grow in both cities. Therefore when planting trees in paved places, or when paving around existing trees, the conditions which a tree needs in order to survive must be considered, and those principles applied to the specifics of each case. If the designer has no particular horticultural expertise, then such advice might well be sought. Different tree species not only grow to different heights with different habits and forms, but also exhibit different environmental preferences and tolerances. It could well be, then, that the appropriate response to specific circumstances is as much a matter of species selection as of the correct treatment of the adjacent pavings.

The principles can be briefly over-simplified thus: a tree gets its nourishment in the form of soluble salts which are drawn in, in water, by the root hairs and carried up by cells in the trunk and branches to the leaves; the bulk of the moisture is evaporated through the leaves (indeed it is this evaporation which, in effect, sucks the water up through the plant) whilst the minerals combine, with sunshine and carbon dioxide, to form starches and sugars, which the tree uses to grow and repair itself. Water is, then, vital to both liberate and transport nutriments. Further, it is important that the soil around the roots should have access to air as well as water. A large percentage of the nitrates absorbed by a tree (and essential for growth) originate in nitrogen in the soil atmosphere. A sealed surface will not only deprive roots of moisture, it will also prevent them from breathing. But a waterlogged soil will also effectively deprive a tree of air, for if the voids in the soil are full of water they cannot have any air. So, whilst the design of a pavement within which trees are to be grown must allow the passage of water of air, it must also ensure that there is not too much water. Ideally, such consideration of the pavement design should extend for the full spread of the mature tree. This is because the roots will spread below ground to much the same extent as do the branches above and it is the roots at the perimeter that are best able to take up nutriments.

One final principle needs to be noted. When paving around an existing tree it is most important neither to raise nor lower existing ground levels. Any adjustment upwards will adversely affect soil air (and possibly water) conditions, and any adjustment downwards will expose the roots themselves.

(i) *Edging pavings adjacent to trees.* Pavings below trees might be treated in a number of different ways so as not to inhibit the passage of air and water. These all have one important common characteristic – they omit the foundation that must be below the paving generally. So there must be a line around each tree at which constructional details change. But this cannot be done by simply omitting the foundation, for that would cause a weakness at that line which, ultimately, must result in the early deterioration of the pavement. A detail must be devised whereby the changes that are necessary either side of that line are accomplished in a way that is satisfactory for the function of the surfacing on **both** sides of it.

Clearly the detail that is appropriate in any particular circumstance will be dependent upon the nature of the paving generally and the nature of the treatment below the tree. But it will also be conditioned by the visual strategy that is to be pursued through the constructional approach to the tree stations. There are two main options. One sets out to make the treatment below the tree as visually similar to the paving generally as is possible. The other exploits the potential differences between these surfaces so as to establish, or augment, rhythm and pattern. Where the first of these options is adopted, the edging to the paving generally should be treated in a way which has minimum visual impact at ground level. Where the second option is found appropriate, the edging itself can be given a role.

(ii) *Surfacing below trees.* Nearly all of the materials which might be employed for pavement surfacing can also be used (with modified details) below trees. The conspicuous exceptions to this list are bound macadams and *in situ* concrete. These two types are essentially impervious to air and water. To the range of materials which might be used, can be added other forms of those materials, and other materials entirely. Those materials which can be used both for paving generally and for surfacing below trees can be used within either a strategy of assimilation or one of contrast. But those forms and materials with applications only below trees must inevitably contrast with the surrounding pavings. In that which follows, materials which have already been discussed for pavings will be considered in sequence, together with illustrative details of how they might be employed. Descriptions of other forms and other materials will then be given, together with figures.

Precast concrete and stone flags. A general area of flags can be carried under a tree provided that they are not butt-jointed and provided that their individual size is not too large. The flags below the trees should be bedded on a minimum of 50 mm (2″) of sand, which is itself laid on a geotextile covering the soil below. The geotextile ensures that the sand is not washed away into the soil. The 10 to 13 mm (⅜ – ½″) joints between the flags should also be filled with sand. One or more flags must be omitted at the tree, and the area immediately adjacent to the trunk filled with loose gravel, or a bark mulch or similar. This approach can be particularly useful in areas with high rainfall and heavy, water retaining subsoils. In such circumstances the pit dug for the tree can act as a sump (resulting in a drowned tree) and so minimum surface porosity can be an advantage. In extreme circumstances the paving may be designed so as to fall away from the tree trunk – all with a view to keeping the subsoil acceptably dry. As the situation becomes less extreme, smaller flags must be used so as to increase the total run of joints and, in consequence, also increase the opportunities for air and water to penetrate. Figure 1.72 shows a detail which gives minimum visual differences between the paving generally and that below the tree.

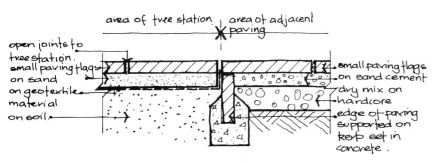

1.72 *This detail shows that there can be little visual difference between an area of paving generally and that below a tree. The edges of the paving generally must be suitably supported and the joints between the flags below the tree left open to allow the passage of water and air. But note that this detail would **not** be satisfactory in areas of low rainfall and/or well drained subsoils.*

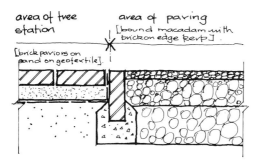

1.73 *Tree situations can be used to generate pattern within an otherwise visually unbroken pavement; this detail uses paving bricks with open joints with an area of bound macadam contained by a brick-on-edge kerb on a concrete bed and haunch.*

Bricks and blocks. These smaller units can be used below a tree exactly as described above. They might, then, be employed when the smallest available precast flags are too large to give an acceptable area of open joint. However, the total visual integration of these with an area of interlocking block or brick paving, is not possible, for interlocking pavings cannot be laid with open joints. Moreover, when the interlocking blocks are non-rectangular, using them with open joints below a tree may prove impossible, for their shape may be such that they can only be butted.

Clearly, when the paving generally is laid with open joints (be that on a flexible or rigid base) then that surfacing can be carried through under a tree with no difficulty. Figure 1.73 shows clay brick pavers laid below a tree as a contrast to bound macadam.

Setts, cobbles and gravel. Stone and precast concrete setts, being very small units, have an essential irregularity which tends to give rather wider joints than do bricks or flags, making them ideal when the intention is visually to assimilate paving below a tree with that used generally. Figure 1.74 shows that, again, the paving generally has to be given a firm edge before the surfacing material is carried through towards the trunk.

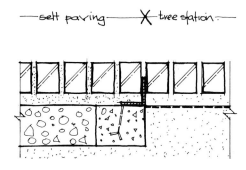

1.74 *Small paving units, such as setts, can be contained with a mild steel frame with fishtailed lugs in a concrete strip. Similar units can then be laid below the tree but with open joints. As there are many more joints, a detail such as this would be much more satisfactory in a dry area than would the arrangement in Fig. 1.72.*

Graded and lightly consolidated gravel pavings for pedestrians only provide the greatest possible opportunity to assimilate the treatment below trees with the general surfacing. The unsealed gravel construction can simply be run up to within about 300 mm (12″) of the tree trunk, with the remaining space filled with the same material carefully tamped by hand (or foot). A geotextile might be used if the gravel has a high proportion of fines. Figure 1.75 shows a detail where stone flags have been used as stepping stones through a surface, which, otherwise, might be rather difficult to walk on.

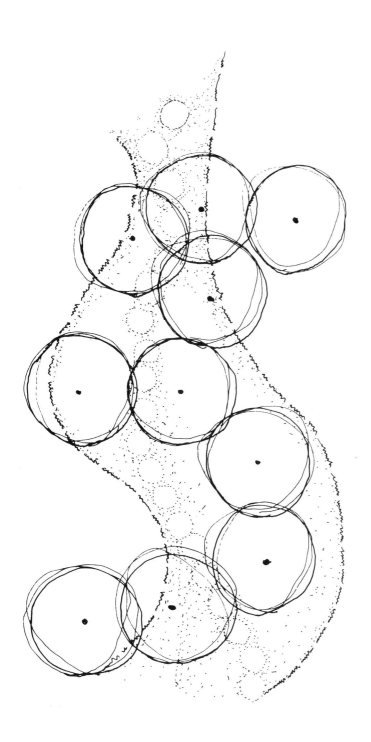

1.75 *Where trees are numerous the special pavings below them may well conjoin. This might be exploited by designers in many different ways. In the example shown here loose gravel is used to ensure that the tree roots get water and air whilst a stepping-stone path of circular flags wind within the gravel and below the trees.*

131

Both graded and single-sized gravels could be used within a general strategy of contrast. Figure 17.6 shows how this might be done with some subtlety, the size and form of the gravel below the tree being exactly that exposed in the *in situ* slab, but of contrasting colour.

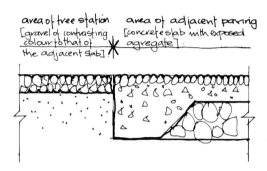

area of tree station [gravel of contrasting colour to that of the adjacent slab]   area of adjacent paving [concrete slab with exposed aggregate].

1.76 *Where there is a high risk of drought, maximum percolation must be allowed below a tree. Loose gravel is ideal in such circumstances. This detail suggests that the loose gravel might be of contrasting colour to that exposed in the neighbouring* in situ *concrete slab.*

(iii) Tree grids. When conditions are such that none of the approaches discussed above would allow sufficient air and/or water to penetrate to the soil below a tree, then some kind of grid will have to be used. These all have a much higher proportion of void to solid than could possibly be the case by leaving open joints between setts, or whatever. Grids of precast concrete and metal are commercially available and timber grids can be constructed *in situ*. Grids fashioned by drilling perforations in stone slabs are a theoretical possibility but their cost would be astronomic.

Precast concrete tree grids. These are manufactured by the same companies which make precast concrete flags. They are all sized so as to co-ordinate dimensionally with particular paving systems. So it is that they are most conveniently used when the pavement generally is precast flags – and of the system of which they are a part. Employed in this way, spaces can be treated as simple compositions of trees (be they formally or informally arranged) over a consistent stretch of paving with no more than a textural change in the appearance of the flags below the trees.

However, precast grids also have potentials within strategies of contrast. Grids of pigmented concrete or with exposed aggregates, possibly picking up the pattern expressed by channels, can be used to highlight an area of plain flags. Alternatively, these or plain finished grid flags can be used with any other pavement structure.

Precast concrete tree grids tend to be rather thicker than a standard flag so as to compensate for the weakening effect of the holes formed in them. Dimensions vary between manufacturers and so trade literature must be consulted. Square, rectangular and hexagonal grid flags are available and, where appropriate, these may have a part-circle omitted to allow construction immediately adjacent to the tree. The perforations which allow the passage of water and air may be circular or slots. Figure 1.77 illustrates two of the range of precast concrete tree grids that are commercially available.

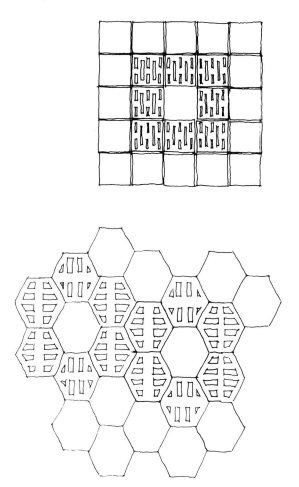

1.77 *Special precast concrete flags are available for use below trees.*

As an alternative to precast grids made specifically for use around trees, the perforated slabs manufactured to reinforce grassed surfaces (as described in section 1.15.7) might be considered.

As before, the paving generally must be suitably edged to ensure its structural validity. And again, that edging might be given a visual impact, or not. The grid flags are bedded on 50 mm (2″) or so of sand on a geotextile laid directly on the soil below. The holes in the grid flags can be traps for litter, and so it is useful to fill them with a small, single-sized gravel.

Metal tree grids. Metal tree grids are made to standard patterns in iron, steel, and aluminium. Other patterns and/or using other materials with sufficient strength (such as bronze) would be possible to special order – at a price. Metal grids have the highest proportion of void to solid of all the alternatives considered here. Moreover, they do not rest on the soil beneath but are supported on a frame, and so provide the highest possible opportunities for the passage of water and air. Some patterns have been devised so as to allow sections to be cut or broken off as the tree trunk grows.

Metal tree grids must inevitably contrast visually with the paving which surrounds them. However, the range of patterns available allows contrasts of greater or lesser degree. All have a robust fitness-for-purpose. Some aim to be no more than that, and so tend to sit undemonstratively around the tree. Others have quite complex patterns and decorations and these can be used to significantly enliven, even dramatize, the floorscape.

As has been said, metal tree grids are supported above the soil around the tree on an independent metal frame. This is usually a standard mild steel angle although other metals (or a galvanized finish) might be specified if the possibility of rusting was a concern. Different manufacturers have different details for sitting the frame on the foundation to the adjacent paving's edging and so catalogues must be consulted. Figure 1.78 shows a frequent approach where fish-tailed lugs are welded on to the frame at intervals, giving each side of the frame at least two points of support. The frame is temporarily supported at the location and level required, and the lugs cast into an *in situ* concrete strip, with the frame also restraining the edge of the general paving.

The space below a metal tree grid can also become a repository for litter. To remove the litter the gratings must be lifted. (Most come with a simple locking device so as to stop them being lifted illegitimately.) This also allows access to the soil should there be a need for weeding or whatever. But if there was a concern that maintenance standards might be low, the possibility of filling the space with unconsolidated gravel might be worth considering – or, a bark or similar mulch might be used which would not only fill the space but also suppress weeds and feed the tree.

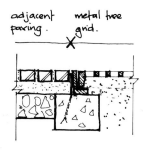

1.78 *Metal tree grids are available in a wealth of shapes and patterns. Fishtailed lugs are welded to the frame of the tree grid and cast into a concrete strip foundation which will also contain the edge of the adjacent paving.*

Timber tree grids. Very little needs to be said here about this possible approach to paving around a tree. The principles of constructing timber decking have been covered in section 1.16. Just such a deck can be used in these circumstances; with a particular value when the tree is existing, for no other kind of tree grid would cause so little disturbance to the ground around the tree. Generally speaking there would have to be a step up from the paving proper to the decking, but it is advisable to minimize the amount of timber buried in the ground. This approach can, however, be most useful when paving levels have to be higher than existing ground. The deck can then be set flush with the adjacent paving whilst levels below the deck remain exactly as before. Figure 1.79 illustrates this notion. With a little

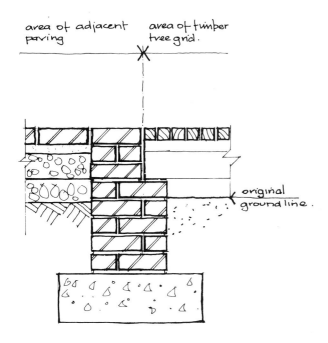

area of adjacent paving

area of timber tree grid.

original ground line.

1.79 *This detail illustrates how a timber tree grid might be used around an existing tree within an area of paving built at a level that is higher than the original ground. Original ground level must be sustained around the tree, or it will die. In this detail the pavings are supported on a low brick retaining wall which also supports the frame to the timber grid.*

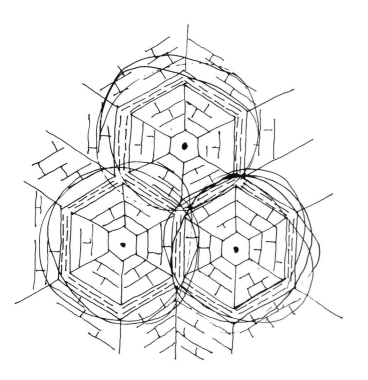

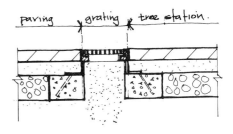

paving   grating   tree station.

1.80 *With very large trees in areas of impervious paving, it can be advisable to allow water and air into the ground below the spread of the tree as well as at its trunk. This detail employs the kind of grating used over linear drains not only to satisfy that function but also to generate pattern in the paving as a whole.*

135

design thought, panels of boards can be made removable by, say, running back half-a-dozen brass screws, so giving access for maintenance.

Perimeter grids. Earlier notes explained that, ideally, tree grids should extend below a tree for the full (and ultimate) spread of its branches. Clearly this is rarely possible. If it were absolutely essential then many of our tree-filled spaces would have to be paved with nothing but tree grids. The problem, in fact, is less of a difficulty with newly planted trees. These, as they grow, can adapt (within reason) to their specific environment. Whilst the fact that they are surrounded by hard paving means that they probably will not grow to exactly the form and stature which might otherwise have been the case, it does not necessarily also mean that they will not thrive. But an existing mature tree would have a much greater difficulty adapting to a changed environment, for it will already have completed the bulk of its growth. Where a timber decking (ideal in such circumstances) would be inappropriate, the inclusion of additional grids at places below the spread of the tree and not only at the trunk, should be seriously considered. Figure 1.80 illustrates how this idea would be used with considerable impact.

Three very simple details supporting, intended as or becoming sculpture.

Chapter two

WALLS AND
FENCES

Buildings excepted, few man-made things have such an effect upon the landscape (country, town or suburb) as walls and fences. One can think of the immense satisfaction given by the humblest of walls; those high, plain brick walls that surround kitchen gardens, or stone walls in hill country, or sea-walls and breakwaters to harbours, or post and rail fencing in the leafy heart of England.

There are also the grander sort of fences round imposing town houses or palace forecourts, in a splendid tracery of wrought or cast iron, and the fine brickwork of eighteenth-century serpentine walls or nineteenth-century railway cuttings. Or, by contrast, the humbler but equally important vernacular that is peculiar to a district and gives a unique character, such as the slate fences in North Wales, or the high banks of the West Country, Lake District, or Cotswolds, or Northumbrian dry stone walling. There is North Oxford suburban, and West London suburban, and Silicon Valley industrial . . . the variety is immense.

The traditional functions of walls and fences are not identical. It is usually a confusion of these functions, coupled with economic stringency, that now causes so many failures whilst answers to these questions were second nature to our forebears. The functions of a wall might be analysed as follows:

1. To form a boundary which will protect property from trespassers and/or keep stock in.
2. To provide shelter from wind and rain.
3. To provide privacy by screening views in, or to capture landscapes by screening parts of the views out.
4. To retain the ground at abrupt changes of level.

A good wall not only fulfils these practical functions, but it also gives an immense psychological sense of protection (through its very solidity?) which no fence can ever do – nor was it traditionally ever the intention that it should.

Good fences will divide property and protect it from trespassers and (apparently paradoxically) provide better protection from wind than can a wall but (with the exception of close-boarded fences and some hurdles) cannot give a psychological sense of protection and privacy – rather, the reverse. A traditional fence, be it the simple white painted rail that discourages motorists from driving into duck ponds, or the railings around Buckingham Palace, is designed to protect as far as is necessary, but no more. The post and single rail is a deterrent to the casual stroller or tired driver, the palace railings are to control vast crowds; neither is intended to give privacy, but to punctuate the view rather than obstruct it, the former with simplicity and the latter with monumentality.

Railings and gates also have the more subtle function of netting the view. Cast and wrought iron railings have done this with complete success in the past. By their own high standard of design they demand attention in themselves, but their real function was not merely to protect property but to lure one on, visually at least, to the half-veiled pleasures beyond the fence. Often when ornamental gates and railings are moved to another site (or more frequently, when the circumstances beyond the railings are radically changed) they lose much of their point, because the new circumstances offer no mystery or surprise, or the contrast between the tracery in the ironwork and massive masonry which flanks it is lost.

Although some very good walling and fencing is now built, it is in a minority. There are also countless examples which, whilst good in themselves, are in the wrong place. On housing estates, for example, back garden fencing is almost always unsatisfactory. Economy forces open fencing instead of solid walling, at least adjacent to the houses. This is due largely to an attitude of mind and nothing can be done unless those concerned believe it important to consider external as well as internal spaces. The Victorians walled the yards and gardens of even their poorest houses. This was a proper decision. Without acceptable levels of privacy, gardens cannot be used as they should be, as extensions to the house.

The space between buildings and the roads on all kinds of housing developments is often (if anything) handled less satisfactorily than the spaces behind. Acres of grass are as monotonous as the deadly acres of paving still found in some slums. When this is seen it is almost always because the house types, their layout, and the treatment of the external spaces, were considered sequentially rather than as a whole. It should not be beyond the capacity of a competent designer to evolve house types which not only satisfy their own concerns but also have potentials to be formed into groups which define and contain usable spaces, with the nature of those uses and the details which support them having informed the thinking about layout and type design before decisions were taken in those respects.

Not that the problem, apparently, goes unrecognized. Attempts are made to compensate for the lack of any sense of place, but with fussy and over-elaborate details (executed, generally, in cheap imitations of the traditional materials) which, in the event, do more to emphasize the absence of unity than to promote it.

Changes of level have immense possibilities as generators of elements which articulate and punctuate the spaces between buildings. Yet these seem, so often, to be treated as problems best swept out of the way. Given that the site is not drawing-board flat (as it never is) so many developers start by sending in the bulldozers to push the soil into unwieldy lumps in those corners where their standard house types cannot conveniently be fitted. This attitude not only denies the individuality of the site and its location, but removes for all time any possibilities of giving the parts of the development their own individuality within the overall theme.

Most non-traditional fencing today is used because it is cheap, not because anybody particularly likes it. This is an appalling state of affairs. Miles of such fencing are erected every year. There are very good examples of metal fences which have been specially designed for schools and factories, but no such attention would seem to have been given to the design of fences for smaller scale spaces. The usual run-of-the-mill stuff is most uninspired, and traditional design largely ignored. It cannot be that the design task would be impossible, so it must be that nobody, as yet, has expressed a particular interest.

Good design need not be expensive. Concrete block walling, for example, has considerable possibilities when simply detailed within the potentials of that material, rather than in a way which is clearly pretentious.

Timber fencing, where it follows traditional patterns, is good. But there is a tendency to transpose vernacular traditions about the country (and the world) and expect them to be in character with their new surroundings. There would appear to be a greater concern to tack on to a social image and its associations, than to exploit the qualities of individual locations. Why it should be that so many people seem to believe that a few metres of horizontally boarded fence, or a gate made out of plastic horse shoes, will give them the lifestyle of a Virginian rancher, remains a mystery. That they can continue with this delusion despite the rumble of the passing juggernauts (an intrusion which might have been significantly reduced through more appropriate design) remains an even greater mystery.

The qualities which we admire in the artefacts of past generations were so often given to them because of severe technological constraints and limited communications. Traditional designs evolved slowly, in pace with evolving practices and patterns. This can no longer be the case. Radically new materials, products and technologies can find their way to every corner of the country in a matter of months, weeks or hours. In the absence of the constraints which guided our forebears, we must add to the criteria which inform our designs.

We must have a better understanding of the materials we employ, a better understanding of the ways we employ them, and a better understanding of the uses and users for which and whom we are designing, than was needed in the past – otherwise we cannot hope to match, let alone extend, the traditions we have inherited. The qualities of fitness-for-purpose, utility, and honesty found in timber and iron fences, and in stone and brick walls, can also be given to fences of plastic or of steel wire, and walls made of concrete blocks, but only if we know what they are and what they might be in their own terms, rather than seeing them only as cheap and insulting second-rate substitutes.

2.1 WALLS GENERALLY

The following notes refer to walls in general. Further reference should be made to the construction notes under each material for details of the particular type.

2.1.1 Foundations

Only the broadest generalizations can be made on foundations; obviously everything depends on the individual site conditions. The object is to transfer the weight of the wall to the ground below in a way which will remain stable throughout the year and throughout the lifetime of the wall. There are three aspects to this. The foundation must have sufficient width to spread the load over an appropriate area of subsoil, sufficient strength to carry those stresses from the wall through to the subsoil and sufficient depth to ensure that it is not disturbed by the actions of frost and/or moisture movement.

(i) *Foundation breadth.* The theoretical minimum breadth for a wall's foundation if it is to support the weight of the wall above is not difficult to calculate. Simple arithmetic will establish the volume per metre (or foot) run of the wall and manufacturers' catalogues will give the weight per cubic metre (or cubic foot) of the materials, and so the total weight per unit length of the wall can be calculated. Knowledge of site conditions and subsoil bearing capacities (see section 1.6.5) will allow more simple arithmetic leading to the required dimension. However, it has to be said that such calculations, more often than not, give answers which seem ludicrously small – thinner than the thickness of the wall itself is not unknown. The loads which come on a wall's foundations are not solely those generated by self-weight. The wind blowing on the wall (and the suction effect as it eddies behind the wall) will have as much, if not greater, effect. The foundation must be able to resist the overturning forces shown diagrammatically in Figure 2.1. The horizontal force of the wind is converted by the construction into a downward force below the lee-side toe of the foundation. Unfortunately, the calculation of the magnitude of these forces is very complex; not so much those within the construction as those which will bear upon the wall. If the wall were out in the open and good meteorological records were available, then a competent structural engineer could complete the necessary calculations in minutes. But the contexts covered by this book are not out in the open. Wind speeds, strengths and directions can be significantly modified by adjacent buildings – and precise anticipations

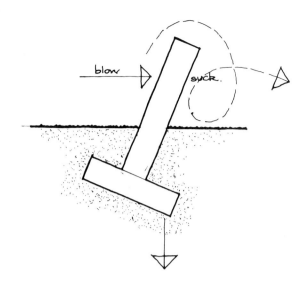

2.1 *A wall does not only have to resist the forces from a wind blowing on it. Eddies on the lee side will also induce suction. These negative forces can sometimes be greater than positive forces to the windward.*

of these effects would probably need a model and a wind tunnel. So, as always, designers must be aware of this possible difficulty and be prepared to seek expert advice when in serious doubt.

Fortunately, the matter is conditioned by one further, and very practical, aspect. It has to be physically possible for work people actually to get (or at least reach) into the trench excavated for the foundation, so as to start building the wall upon it. 450 mm (18") is, then, a practical minimum width for any wall's foundation, which might be increased to 525 mm (21") or 600 mm (24") with increases in the height and/or width of the wall. (As an aside, mechanical diggers come equipped with interchangeable 450 and 600 mm (18 and 24") but rarely 525 mm (21") buckets.)

(ii) Foundation strength. This has two aspects. The strength of the concrete itself, and the thickness of the foundation. As with establishing breadth, theoretical calculations will generally show that the concrete for a wall's foundation needs little strength in itself. If hand-batched a 1:3:6 mix will almost always be adequate, or, if pre-mixed, a concrete giving 14 N/m^2 (2 000 lb/in^2) after 7 days. The thickness of the foundation is more important. The overturning effects of the wind will put bending stresses into the foundation when forces below the lee-side toe are resisted. Unreinforced concrete performs poorly in bending, and the thinner the slab the more poorly it will perform. A good rule-of-thumb is to give the foundation a thickness that is at least equal to the breadth of its shoulder – that is the dimension from the face of the wall to the edge of the foundation. For example, a 600 mm (24") wide foundation to a 225 mm (9") wall should be given a thickness of 187.5 mm, or 200 mm (8") as a more convenient round figure (Figure 2.2).

143

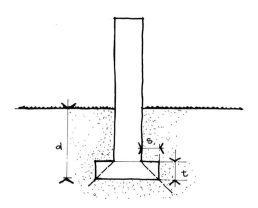

2.2 *The foundations to a free-standing wall are almost invariably over-designed in terms of the vertical loads imposed upon them. Overturning is a greater concern. Their thickness t should be at least 100 mm (4"). The amount that the foundation projects beyond the face of the wall on each side s should be at least equal to the thickness, so that the spread of forces is contained within the width of the foundation as a whole. Overall depth d will vary with subsoil and climatic conditions.*

(iii) *Foundation depth.* This too is conditioned by two aspects. The likely penetration of any frost, and the likely behaviour of the water-table. As is well known, the volume of a unit of water is less than that same unit when frozen. So waterlogged ground will expand as it freezes. If such expansion happens below a foundation, then both it and the wall above will be disturbed. Whilst it might not happen that the wall immediately falls over, it will certainly be weakened, with an increasing likelihood that it will someday fall over, perhaps during the next wind-storm. The problem is easily avoided. Rarely will frosts in Britain penetrate more than 450 to 600 mm (18–24") so, if the underside of the foundation is at that depth or greater, there will be no difficulties in this respect. If building in a more exposed part of Britain (or in another country) then local advice and records should be consulted.

Coping with a variable water-table can be more complex. The difficulty which has to be avoided is not dissimilar to the problem of frost heave. Some subsoils (heavy clays are notorious in this respect) expand and contract as they wet and dry. In such circumstances, stability will be achieved if the soil immediately below the foundation is always wet or always dry, but not if it varies. Trial holes should be excavated to well below the depth at which ground water is found at that time. Changes in the colour of the clay will indicate water-table variations. Again expert local advice should be sought if faced with uncertainty. Trenches for wall foundations should be excavated to a depth that is lower than the lowest anticipated water-table level. Remember also that trees (particularly certain species such as poplars and willows) take vast quantities of water out of the ground as they transpire. Felling existing trees or planting new trees, might significantly change the situation evidenced by the trial holes. Whilst it might seem excessive, it is worth while taking a wall's foundations down to well below the depth of neighbouring roots in such circumstances. When that is done, wall and tree can survive in mutual harmony.

144

Two final points need to be made. First, little of the above has any relevance when building on solid rock. This itself will provide a more than adequate foundation. The concrete which then needs to be introduced into the trench need do no more than make the surface flat and level to facilitate brick and block laying. Secondly, when building on a sloping site the wall foundations must themselves step so as to avoid excessive excavation and construction. The height and length of step should be judged in the light of the size(s) of the material(s) with which the wall is to be constructed as well as the angle of slope. Nowhere should the depth to the underside of the foundation be allowed to be less than the minimum necessary and the upper length of foundation should be allowed to overlap with the lower length at each step for a distance equal to the width of the strip, as ilustrated in Figure 2.3.

2.1.2 Damp-proof courses

It is advisable to keep a free-standing wall of brick, concrete block, dressed stone or the like, as dry as possible. The water can cause both chemical and mechanical damage – the more so if it freezes. Minimizing the penetration of rain-water is considered below. But damp can also rise up out of the ground through capillary action within the wall itself. So, a dpc (damp-proof course) is frequently included within the wall's construction. A wide range of materials are available for this purpose, their one common quality being that they are all impervious to water. All should be placed never less than, but also not much higher than, 150 mm (6″) above the adjacent ground surface. If lower than this, there is a real risk that heavy raindrops will bounce up to soak the wall above the dpc to a greater extent than would the rising damp. But if much higher, the advantages of including the dpc will be largely lost. Some dpc materials are built into one or other of the lower horizontal bed joints in the wall; others substitute for the whole of one or more of the lower courses.

(i) *Sheet metal*. Lead, copper, or other sheet metal that will not rust away, but with sufficient malleability not to form a complete discontinuity within the construction, can be used in strips the width of, or very fractionally narrower than, the wall. They are laid immediately on the upper surface of the course below, with the bedding mortar for the course above laid directly on them. They are very expensive.

(ii) *Asphalt*. A course of mastic asphalt can be substituted for the mortar bed at an appropriate level. This would be a very inconvenient detail unless mastic asphalt was being used for some other purpose on the site.

(iii) *Impregnated felts*. Bitumen or pitch impregnated hessians and felts, or their synthetic equivalents, are probably the dpc materials

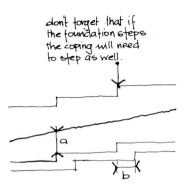

don't forget that if the foundation steps the coping will need to step as well.

2.3 *When a wall is constructed on sloping ground its foundation will have to step. Care must be taken that the foundation is never at less than its minimum depth a. At each step the foundation should be twice as thick as it is elsewhere for a length equivalent to its width b.*

most frequently employed in the building trade generally. They are available in rolls of various widths which co-ordinate dimensionally with the various widths of wall that can be constructed with bricks and concrete blocks. They are used in the way described above for a metal dpc. These are economical both as materials and in terms of associated labour costs – particularly because that labour is likely to be very familiar with them.

(iv) *Slate.* Slate has a long history of use as a dpc. Whilst a single slate is impervious to water, each is of relatively short length. So, one layer is thinly bedded on mortar on the wall below, with a second layer thinly bedded on that, in such a way that the gap between any two slates is immediately above the centre of the slate below. This gives any damp which tries to find its way through the mortar a very long path to travel.

(v) *Engineering bricks.* Engineering bricks are also impervious to water, but their length is even less than that of slates. For this reason, principally, they are usually used in two courses (simply substituting for the bricks or blocks), in order to more than treble the distance damp would have to travel, compared with the length of a single perpend.

Figures 2.4 and 2.5 illustrate typical ways in which these different dpc materials can be used.

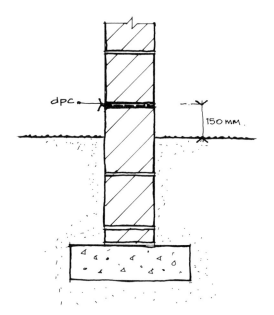

2.4 *The damp proof course (dpc) in a free-standing wall should be at least 150 mm (6") above the ground but never much more.*

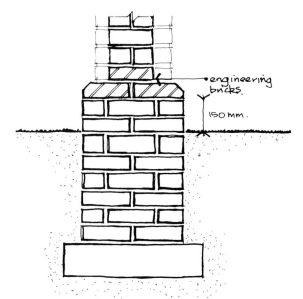

2.5 *A dpc of two courses of engineering bricks does not interfere with structural validity of the wall. Further, colour and shape contrasts can be exploited visually.*

146

The dpc detail can also have an effect on appearance. This is much less so with metal or impregnated felt materials – the most that is likely to be seen is a slightly thicker bed joint towards the bottom of the wall. A slate dpc results in an appreciably thicker bed joint with, if desired, two neat thin lines expressed within it. But clearly it would be an engineering brick detail which would have the greatest visual consequence. It is most likely that there will be both colour and textural differences between the engineering bricks and the materials used in the wall generally. With care this can be most attractive, a quality which can be amplified by the use of similar materials at other vulnerable parts of the wall.

2.1.3   Height to thickness ratios

As is so often the case, the solution to one problem itself creates other problems. This is so with (some) damp-proof courses. The ways in which the wind might attempt to overturn a wall was described when considering foundation design. But having an adequate foundation would be no comfort whatsoever if the wall should blow over because that part above the dpc becomes detached from that below. This can happen all too easily with metal and bitumen impregnated (and similar) dpc materials. The adhesion between the mortar and the dpc will, at best, be poor. The situation is a little better with a slate dpc which does have a minimal mortar bed both above and below. But it is only with an engineering brick dpc that one can say with some certainty that there is no discontinuity at dpc level. However, this does not mean that an engineering brick dpc is the only truly

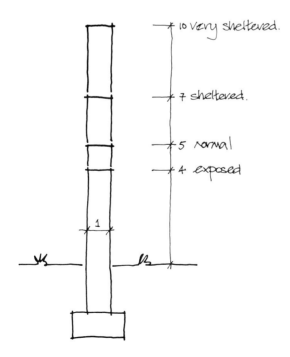

2.6 *Brick, block or stone walls should have their height limited to the above mutiples of their width, in variously exposed (to wind) locations.*

147

acceptable option. The wall itself can be designed to give it stability through weight, thickness, and/or buttressing.

In very sheltered conditions, a maximum height:thickness ratio of 10:1 is recommended; in sheltered conditions this is better at 7:1; in more normal conditions a ratio of 5:1 is more satisfactory; whilst in exposed areas the height of a wall should be limited to 4 times its thickness. These recommendations would give some very thick walls in certain circumstances, if that thickness was consistent along its whole length. But the wall can be buttressed and if it has these proportions at the buttresses and these are at sufficiently close centres, then the wall will be sound. Figure 2.6 suggests a rule of thumb for buttress centres.

2.1.4 Mortar for masonry

Mortar, whether for brick, stone or concrete blocks, must be chosen with great care, since the nature of the joints can have as great an effect upon the appearance of the wall as does the basic material. Allowance should be made, in the specification, for trial panels to be set up, and the contractor should be told at the tendering stage if special ingredients (such as particular sands) or additives are, or are not, to be used. Trial panels should not be inspected before the mortar has had time to dry out.

(i) Strength of mortar. Mortar has been defined as a gap-filling adhesive. The bulk of any mortar is sand, and with this is lime (non-hydraulic or semi-hydraulic) and/or cement. Lime mortars stiffen slowly as they dry out whilst cement mortars stiffen through a chemical reaction and so gain strength more quickly, and can do so when wet. Lime mortars, then, remain vulnerable to frost attack for several months; and they never gain the same degree of strength as do cement mortars.

The highest possible mortar strength is not required in every circumstance. Ideally, it should be very slightly weaker than the basic material of which the wall is built. Too strong a mortar will make a very rigid wall with no elasticity in the joints to take up expansion and contraction movements with changes of temperature. This would mean that instead of many minute hair-line cracks developing through-out the wall, it is likely to crack in just a few places, with those cracks being large. Further, a very strong mortar could accelerate the weathering of a relatively soft basic material. The face of a wall must inevitably become soaked by rain-water. Strong mortars will concen-trate that water in the basic material. Wet materials are more vulnerable to frost attack, and materials which are constantly wetting and drying are more likely to suffer chemical changes. If the basic material weathers faster than the mortar, the joints will start to become proud of the face of the wall, and the conditions described immediately above made worse.

148

The mortar mixes recommended for free-standing walls vary, then, with basic materials. Some examples are given below. Ingredients are given as the ratios of cement:lime:sand.

1. Clay, calcium silicate and concrete bricks on sheltered sites or built in the summer – 1:2:9.
2. Clay, calcium silicate and concrete bricks on normal sites or built in the winter – 1:1:6.
3. All bricks, below dpc, at copings, or in retaining walls – 1:0:4.*
4. Dressed stone – 1:3:10 to 1:4:16 depending on softness of stone.
5. Concrete blocks generally – 1:1:6.
6. Concrete blocks below dpc, at copings and in retaining walls – 1:0:3.*
7. Clay engineering bricks – 1:0:3.*

(* When no lime is to be included in a mortar it is usually specified simply as the ratio of cement to sand, i.e. 1:4 or 1:3.)

(ii) *Mortar materials.* Cements and building limes, and sands for mortars, are covered by a number of BS specifications (given in Appendix A). Sands for mortars include both naturally occurring sands and sands made by crushing stone. These last are of particular value when the basic material is stone, for the sand can be made from the same stone, giving both structural and visual consistency. The water must be clean and free from disadvantageous chemicals in solution. Water fit for drinking or from a potable source are traditional specifications.

Bricklayers and masons, understandably, prefer to work with a mortar that retains its workability and does not start to stiffen too soon. Mortars with a lime content have this quality. So do masonry cements which have a proportion of inert filler, such as ground chalk. (When using a masonry cement the amount of cement in the ratios given above should be increased by 30%. That is, a 1:4 ordinary portland cement:sand mortar would have similar strength to a 1:3 mortar using masonry cement.) Other additives, usually called plasticizers can be added to mortars to give workability. These should always be used as specified by their manufacturer. But it should be noted that these agents will slightly change the colour of the mortar. It might be that the designer prefers to specify that they should not be used. It should always be stated that if they are to be used, that should be all the time, and not only in hot weather when they are particularly useful.

Walls cannot be built during frosty weather, for if the water in the mortar freezes it will not be available to the chemical reaction (see also notes on concrete laying, section 1.13). It is often specified that wall construction should cease if the temperature falls below 2°C (35°F) and not start again until it rises to 0°C (32°F). Agents can be added to mortar that will allow finer tolerances than these. Again, the manufacturers' specifications should always be followed and it should be specified that, if they are to be used at all, they should be used all of the time and not only when it is cold; otherwise mortar colours will be patchy.

Some bricks have relatively high proportions of sulphates, as do some subsoils. This can cause difficulties in saturated conditions. Sulphate-resistant cement should be used in such circumstances. It is used in the same proportions as ordinary cement but, yet again, its consistent use would be necessary to ensure consistency of appearance.

Chemicals are available, both as liquids and powders, which can be added when mixing mortar in order to give it waterproof qualities. None achieve total impermeability (although they can do when used in a much thicker concrete slab) but will significantly improve performance in this respect. The inclusion of such additives in the mortar used at a slate or engineering brick dpc or below copings, would undoubtedly improve those details. Because there will already be visual changes at these places within the wall, it would not be necessary to require the additive to be used in all of the mortar.

(iii) Joints. The way in which mortar joints are finished off is important to both the appearance and durability of the wall. The general objective is to minimize the opportunities for rain-water to lodge at the joints, and thus soak into the wall. But, a simple flush finish can look most unsatisfactory, particularly when the basic material has some texture. Figure 2.7 shows a range of ways in which joints might be finished, some introducing shadow lines – which can do much to enliven textural qualities.

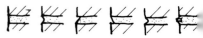

2.7 *The mortar joints in masonry can be finished in a number of different ways. From left to right – flush, bucket-handle, weathered (correct), weathered (incorrect), keyed, and tucked pointed.*

It can be that a quality of finish is needed for the joints that would be most expensive if the same mortar had to be used throughout the wall. This is most frequently the case with ashlar and other dressed stone walls. The general mortar is then kept back from the face of the joint as the wall is built, with all of the joints being pointed with a special mortar as a finishing operation.

2.1.5 Copings

The function of a coping is to prevent water penetrating the top of the wall, and to throw the water which collects on the coping well clear of the face of the wall. It is rather like a hat sitting on top of the wall and, like a hat, its function does not stop at keeping the weather out. A coping has an immense effect upon the character of the wall it caps; it may be a crisp line, a mossy slab, or a crude lump of concrete. Or (and equally important) the wall might be left bare-headed, with no obviously separate coping at all; simply the basic material arranged a little differently.

Generally speaking it is most satisfactory to give the coping a projection of 38 to 50 mm (1½" – 2") to ensure that the water is shed well clear of the wall's faces. Where the nature of the materials allow, it is also best if the underside of the projection is given a narrow rebate to make sure that the water does drip off. Clearly, such a detail cannot act as an umbrella keeping the whole of the wall dry. That is not the intention. What the coping does is keep the top of the wall at least as dry as the rest. It is saturation through the whole thickness of the wall that is to be avoided.

Some coping materials are themselves impervious to water. Others are not. In these latter cases, a dpc must be included below the coping. The whole range of dpc materials discussed in section 2.1.2 could also be used, suitably detailed, below a coping.

(i) Stone copings (other than slate). Stone is expensive, but there are cases where an existing tradition is to be extended, or old work is to be repaired, or where the solid and constant qualities of stone can be exploited, which make the cost justifiable.

The stone used should be of low permeability and resistant to freezing. Limestones are generally good, but sandstones may shrink and leave fine cracks at the joints (possibly unfortunate in carefully dressed work). There is a BS specification for stone copings (given in Appendix A). This covers quality of stone, and states that the copings should be cut so that they can be laid on their natural bed, unless the quarry advises otherwise. Unless a particular finish is specified, they are dressed with rubbed or fine sawn surfaces. Lengths are given as not less then 600 mm (2'0") nor more than 1200 mm (4'0"). This is a compromise between wanting as few joints as possible and being able to handle copings without damage. Widths are given as between 300 and 350 mm (12–14") as specified. Both saddle-backed and splayed copings are illustrated. Standard specials include stopped ends (where the section simply stops with a dressed face); returned stopped ends (where the section is returned across the end); and angled returns of 90, 120 and 135° (which allow the coping to be taken around a corner in the wall).

It must be added, however, that stone copings are no longer commonly used, so, when they are, the designer should be able to specify whatever is wanted at no significant increase in cost over items included in the BS. These stone copings are not impervious to water and so must be used with an associated dpc. The simplest detail is to lay an appropriate width of sheet dpc on the top of the wall proper, and then bed the coping on mortar on top of that. Figure 2.8 illustrates such a detail.

(ii) Slate copings. Most slates are (for all practical purposes) impervious to water even when split very thin. So, the fine line given by a slate coping can look very trim, allow for very fine jointing, and need no additional dpc. But as with other stone copings, they are not cheap. Precise dimensions vary between quarries. Thicknesses vary between 25 and 50 mm (1–2") and sections may be weathered or flat. Lengths of up to 2 100 mm (7'0") are obtainable but 1 200 mm (4'0") is a more normal maximum. Random lengths with a minimum of 600 mm (2'0") are frequently specified and, because this gives the

2.8 *All free-standing walls must have a coping – a weatherproof hat. This example in dressed stone is but one of a number of sections which might be used. Note that drips should be included and the coping set on a dpc.*

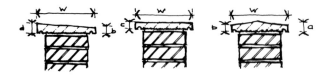

2.9 *Three examples of slate copings. Overall width w should be the wall thickness plus 75 mm (3"). Slate allows the coping to be expressed as a very thin line. Maximum thickness, a and d can be between 38 and 50 mm (1½ – 2"). Minimum thickness, b and c, can be 25 to 38 mm (1 – 1½"). Note that a slate coping does not need a dpc, is keyed on its underside to give grip with the mortar bed, and induces drips.*

quarry some flexibility, reduces costs. Figure 2.9 shows some typical details.

Coping slates have a scored under-surface to provide a key, and should be bedded on a 1:3 cement:sand mortar directly on to the top of the wall. The individual lengths of coping can be dowelled together using slate or copper pegs, and/or fishtailed non-ferrous cramps may be used to ensure the coping cannot lift from the wall. Fine rubbed, frame-sawn and naturally riven finishes are all available.

Slate quarries in this country no longer turn out the millions of items each year as was once the case. Decreasing demand has led to increased costs. But when costs are high the designer can afford to specify more precisely. So, whilst slate copings are still produced as standard items (refer to manufacturers' catalogues) it could be well worth while making further inquiries should the standard not be exactly what is wanted.

There is one particular disadvantage of slate. The way in which it was formed, under heat and pressure, resulted in the formation of planes of crystals within its structure. It can be easily split between those planes. It is this quality which led to its almost universal use as a roofing material before the technology for mass producing concrete tiles was developed. But these planes are still within the material when it is split or sawn to larger thicknesses. It can be that they are opened up by frost. The quarrymen have the skills necessary to recognize slate that is less likely to behave in this way. Designers must make sure that satisfactory material is both specified and supplied.

(iii) Precast concrete and reconstructed stone. Technically speaking, there are no differences between precast concrete and reconstructed stone. The latter is simply the former but made with greater care for materials and moulds.

Precast concrete copings are not totally impervious to water, particularly as they tend to open at the joints, and so must be laid on a dpc in one of the ways described for stone in section 2.1.5(i). They should be bedded on the dpc with a cement-sand mortar as used in the wall generally, and laid with a 10 to 13 mm (⅜–½") joint. The joint should also be filled with mortar. Some coping units have a frog (or depression) cast into their end faces so as to provide a better key for the joint mortar. (Joints without a frog have to be specified for the ends of the wall.)

There is a BS specification for precast concrete copings (given in Appendix A). This covers materials, curing, finishes and forms.

Three patterns are suggested with widths of 300 mm (12″) for 225 mm (9″) walls, and 350 mm (14″) for 275 mm (11″) cavity walls and for 225 mm (9″) walls that are to be rendered. Stopped ends, stopped return ends, and angled returns of 90, 120 and 135° are included. The value of this BS, or more precisely, its coverage of patterns and sizes, can be debated. First, the visual weight of a coping which looks right when above the eye, is not the same as that which is appropriate on a low wall. The proportions of the wall as a whole are different, and looking up one sees much less of the coping than when looking down. Yet the BS does not include sections with the necessary variety. Secondly, the production costs of precast concrete copings is very low. If the designer has a need for a large number of copings, then they can be mass-produced to an individual design at little extra cost. If there is a need for only a few, then they can be cast on site very cheaply and effectively. Readers will already have appreciated that the authors see considerable potential in concrete as a landscape material, but only if it is used in a way which expresses its own qualities and is not viewed as a cheap substitute for some other material. The BS for precast concrete copings has many similarities with that for natural stone copings. The materials are not the same and should not be considered as such.

When mass-produced, these copings are usually made in steel moulds. The notes for precast concrete flags in section 1.15.3 are relevant here also. When cast on site (but not necessarily *in situ*) the notes for concrete walls and concrete finishes, (sections 2.3 and 1.13.8) are germane.

Finally, most of the manufacturers of precast concrete copings make a much wider range than covered by the BS. Reference to their catalogues would be useful when circumstances demand off-the-peg items.

(iv) Clayware copings. These are made of the same materials, and in essentially the same ways, as are bricks. Indeed some, in effect, are specially shaped bricks. Others have an overall length of 215 mm (8½″), for example, allowing their joints to align with alternate perpends in a brick wall below. Again there is a BS specification of dubious value once it extends beyond describing the materials. This seems to have been recognized by manufacturers who rarely include BS sections in their catalogues of standard products.

Low permeability is essential – much lower, possibly, than that of the bricks in the wall below. But, because there will be a relatively large number of joints, the inclusion of a dpc is advisable.

Both glazed and unglazed copings are available. The natural unglazed colours of the fired clay (generally deep reds, browns, purple-reds and blues) have much to commend them, whilst some of the glazed colours have higher chroma with a value when the wall needs to be given a visual dominance.

The materials are variously described as clayware, terra cotta, stoneware and faience, depending upon the precise nature of the basic clay and the way in which it has been fired. There would appear to be an increasing interest in the production of these copings to the patterns of the past.

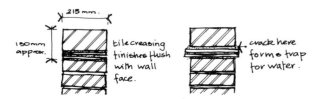

2.10 *Brick-on-edge copings can look very neat and tidy. A tile or roofing slate creasing provides a dpc. The creasing can be projected to provide a drip but this can lead to other problems.*

(v) *Brick-on-edge copings.* It is dimensionally very convenient to finish the top of a 215 mm (8½″) thick brick wall with the same bricks as are used in the wall but laid with their stretcher face uppermost. There are also visual arguments for this detail when it is the qualities of the wall as a whole, rather than of its parts, that are to be stressed. But there are also disadvantages. The brick in the wall below may have too much porosity for use as a coping. Special quality bricks (not to be confused with special shapes) are needed for a coping. The detail exposes many joints to the weather above. A dpc or slate or tile creasing will be needed. (Figure 2.10 illustrates a tile creasing.) The top of the coping will be flat. Even a special quality brick may suffer from this treatment, and the mortar in the joints almost certainly will. The bed face of the bricks will be exposed to view at the ends of the wall. These are rarely very decorative, and when they are not unattractive in themselves, are often significantly different in appearance to the other faces. Worst of all, many bricks have lightening holes within them (which also ensure even firing within the kiln). These not only look unattractive but conduct rainwater right into the heart of the coping. Finally, the last brick at the end of the coping, will be supported at one side only. It will take only a few children (running on and jumping from the wall) to dislodge it – and if there are no children, the frost is likely to have the same effect. So, a simple brick-on-edge coping could only be recommended when budgets were very low.

Fortunately, the brick manufacturers are well aware of the defects of the simple detail. Special shapes (as well as special quality) are offered as standard, some specifically for use in this context, which will effectively shed water. Bricks with the same finish on one bed face, as is on one stretcher and the two header faces, are available. Blocks with an overall size equivalent to three bricks on edge, can be used to give stability at the ends of the wall (and some of these have dummy joints which can be pointed on site with the mortar used generally). At least one manufacturer produces a system for brick-on-edge copings which not only counters these defects but also incorporates the dpc. So, the elegant simplicity of brick-on-edge copings can be achieved and without technical weaknesses, through appropriate use of specials where necessary.

(vi) Roofing-tile copings. The imaginative designer can always find ways of using products made specifically for one purpose in order to resolve a quite different problem. Items made for the roofs and ridges of buildings also have potentials for the copings of free-standing walls. One such detail is illustrated in Figure 2.11. The possibilities are too numerous to allow comprehensive illustration. But two points must be borne in mind. A high wall will expose to view the underside of a projecting coping. These may be parts of a roofing tile that the eye was never intended to reach. And many roofing tiles stay in place on a roof due to their collective weight. Simply bedding them on mortar as a coping may not be sufficient, particularly if vandalism (deliberate or accidental) is likely to be a problem.

(vii) Metal copings. Preformed copings of copper, zinc, aluminium and stainless steel, plus some alloys could be considered. They will give a fine trim-line to the top of a wall which can be of value when neighbouring buildings are roofed in that metal. They are particularly useful on rendered walls where a totally impervious coping is essential if the weather is not to penetrate between the render and its background.

Where a metal coping is to be used in conjunction with a metal roof, the roofing material will probably dictate the coping metal; not only for visual reasons but because the possibility of electrolytic action is to be avoided. Whilst this is less essential with an impervious coping, it remains best if a metal coping is given a slight fall to one or other side. No flat surface is ever entirely flat. Puddles on a low wall will look unsightly (and ruin its possible use as an impromptu seat), and evaporating puddles will cause staining, if not worse. All should project at least 25 mm (1″) at each side and be formed in a way which provides drips. Fixing and jointing should allow for thermal movement when lengths exceed 9 m (30′0″). And the possibility of electrolytic action must be avoided by the use of correct screws and other fixings. Figure 2.12 illustrates a range of details recommended for different materials.

2.1.6 Movement joints

A free-standing wall is liable to expand and contract with the seasons, and the initial shrinkage of walls built wet can also present problems. Their behaviour is difficult to predict with any certainty because so much depends upon the precise site conditions. Under otherwise similar conditions, an east–west wall may behave in an entirely different way to one which runs north–south. The use of mortar of appropriate strength has already been discussed in this respect (section 2.1.4). But with extensive areas, lengths, and sometimes

2.11 *Free-standing walls might be given a match with the buildings they adjoin through the use of roofing and ridge tiles as their coping. Details will vary with the particular items used, but remember that those details will almost certainly not be the same as when they are used on a roof.*

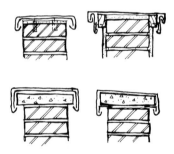

2.12 *Pressed metal copings can be made from aluminium, zinc or copper. Details vary with metals, manufacturers, and the materials of the wall. Fixings might be with clips built into the wall (top left), to hardwood battens themselves fixed to plugs in the wall (top right), or by mechanical fixing to a concrete sub-coping (lower left and right).*

masses of wall, special joints which allow expansion and contraction must be included. The ways in which these are detailed, and the intervals at which they should occur, varies with materials, and so are covered individually in the sections which follow. However, one general point might be made. Well designed details always have a quality of self-confidence – they are not diffident about being as they have to be. This is so however humble or grand their character. A movement joint included in a wall with no more thought as to its location than the intervals at which such joints should occur, could never have such a quality. A good designer will have checked on those intervals at an early stage, and applied that dimensional discipline to the design as a whole. The positions of pillars or buttresses, of panels in the face or of offsets in the line of the wall, can all allow movement joints to be proud of their necessary existence.

2.1.7 Junctions between free-standing walls and buildings

It is inevitable that there will be differential movement between a free-standing wall and one which is a part of a building. There can be no doubt that the technically correct detail in such a circumstance is to leave a mortar joint thickness gap between the two (which might then be treated as an expansion/contraction joint). However, there is also no doubt that this can significantly disrupt what might otherwise have been a simple and successful integration of internal and external spaces. It is imperative that the building and landscape designers (where they are not the same) get together to resolve this detail to their mutual satisfaction. Approaches of the kind discussed for movement joints generally might be considered. Alternatively, it might prove possible to allow for movement a little way away from the building, with the initial length of the free-standing wall being built as a part of the building contract.

2.1.8 Buttresses

As has been explained, the force which is most likely to overturn a

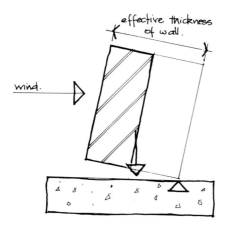

2.13 *Buttresses in a wall increase its effective thickness and thus its resistance to overturning.*

soundly built free-standing wall is the wind. This force can be resisted by sheer thickness and weight. But, with a tall wall this would result in the use of a great deal of material which could prove very expensive. A wall can be made so that it will resist overturning by incorporating piers or buttresses at intervals. That is, the forces on the panels of wall between the piers are transferred to the piers which, themselves, have sufficient weight and thickness to resist those loads. Figure 2.13 illustrates these points.

A buttress is a short fin, of the same material as the wall generally, which projects on one side at right angles to the general line. The buttress might project further at the base of the wall than it does near the top, for it is thickness at the base which does most to resist overturning. A pier is essentially similar to a buttress but projects on both sides of the wall. A pier would be more valuable than a buttress when strong winds might come from any direction, and/or when the same appearance is wanted on both sides of the wall. The eighteenth-century serpentine walls, which had remarkably little thickness for their height, needed no buttresses or piers because they were, in effect, a continuous sequence of curvilinear buttresses.

Similar effects can be achieved with straight-line geometries. Figure 2.14 illustrates some approaches in principle. But it should be noted that all of these, and others which might have been illustrated, will radically change the appearance of the wall. If a free-flowing and essentially horizontal line is required, then stability must be given through thickness.

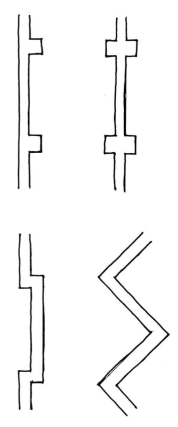

2.14 *A wall might be buttressed in a number of different ways, with each having different visual consequences.*

2.2 BRICK AND BLOCK WALLS

Walls might be constructed from bricks made of clay, calcium silicates (flint–lime and sand–lime), or concrete. These are all made to a format size of 225 × 112.5 × 75 mm (9 × 4½ × 3″). The format size is the size of the brick itself plus an allowance for jointing. Some tolerance has to be allowed for the size of individual bricks, but for design purposes this is normally taken to be 215 × 102.5 × 65 mm (8½ × 4⅛ × 2⅝″). This size and form is by no means accidental. It has developed over the centuries to give bricks a size and weight that can be picked up in one hand by a bricklayer who needs to have the other hand free for the trowel.

It is also a miracle of simple dimensional co-ordination. Its length is twice its width and three times its height. This allows bricks to be organized in space in a wide range of ways giving necessary structural performance and a choice of textural appearance. These textures or patterns on the face of a wall are the result of bonding the bricks, as they are laid, so that no two immediately adjacent perpends line through. The perpends are the vertical (perpendicular) joints between the bricks. If these were to line through, much of the potential strength of the wall as a whole would be lost. Figure 2.15 illustrates this.

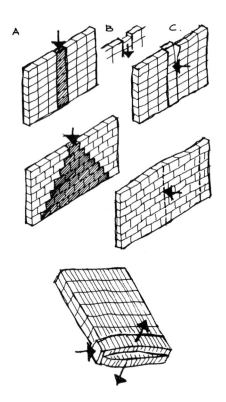

2.15 *Masonry walls must be bonded if vertical (A and B) and horizontal (C) forces are to be resisted by panels of masonry rather than columns of individual units. Walls built as two skins must be bonded (or incorporate wall ties) if those two skins are to act as one.*

There are a number of ways in which a wall might be bonded. These will give different degrees of structural strength. However, for a free-standing wall those differences are not likely to be significant and it is more likely that the designer will select the intended bond for visual rather than practical reasons. Some examples of common forms of bonding are shown in Figure 2.16, but these are by no means all.

There is no doubt that the skills to design and build interesting brickwork fell into deep decline during the not too distant past. The most plausible explanation for this is the almost universal introduction of cavity walls for buildings with brick external walls. The half-brick thick external leaf to such walls can only be economically built in stretcher bond. One suspects that there must be bricklayers who have never built anything else. Stretcher bond is the blandest of all the alternatives. Whilst its simplicity might be found refreshing in an overly complex situation, it does become deadly dull when that is all that is seen. However, there does seem to be an awakening interest in architectural brickwork. Let us hope that this soon extends into the external spaces as well.

Fortunately, straight half-brick thick walls have little value as free-standing walls. If much more than two or three courses high, they would soon get knocked over. This weakness can be offset through buttressing, with piers, by stepping or curving the wall to incorporate

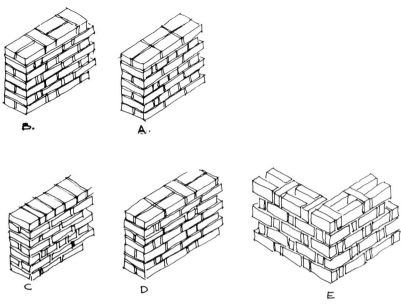

2.16 *Different approaches to bonding have different visual effects. The examples here are English bond (A), alternate courses of headers and stretchers; Flemish bond (B), headers and stretchers alternate in each course; English garden wall bond (C), three courses of stretchers above each course of headers; Flemish garden wall bond (D), three stretchers between headers on each course; Rat-trap bond (E), which exposes what would normally be the bed face of the brick and contains voids within the wall, in order to save weight and material.*

buttresses, or by using a thicker wall with a visually appropriate bond. So it should be very difficult to build an uninteresting free-standing brick wall. Why then have so many designers managed to achieve just that?

2.2.1 Clay bricks

Clay bricks are classified by variety and by quality. **Common, facing,** and **engineering** are varieties of clay bricks. **Internal, ordinary,** and **special** are of different qualities. (But do not confuse special quality with special shapes – they are not the same thing.) A free-standing wall might be built of any **variety** of brick – although if a common, the wall would not look very pretty. Common bricks are normally used when the wall is to be rendered or covered over in some other way. But only special **quality** clay bricks are suitable for external free-standing walls. There is, in particular, a need for good frost resistance, for a free-standing wall will get more thoroughly wetted than will that to a building, and so will be more vulnerable if frozen when wet. No entirely satisfactory test for the frost resistance of clay bricks has ever been developed. Whilst it is known that high strength and low water absorption properties do lead to good frost resistance, it is also known that there are bricks which cannot be described in those ways which are equally satisfactory. Where direct experience is not available the manufacturer should be asked to present evidence of the successful use of a particular brick in a location with similar or greater degrees of exposure to rain and frost.

Clay bricks are also more vulnerable to sulphate attack when in a free-standing wall than when they are part of a building. Again this is because they are likely to be more thoroughly wetted. Further, the cement in the mortar of a free-standing wall is also more vulnerable to such attack. Clay bricks with a sulphate content of below 1% are recommended for free-standing walls whilst 3% is usually tolerated for buildings.

Clay bricks can contain salts other than sulphates. These cause no physical damage but can disfigure the appearance. The salts are dissolved when the wall is wetted and are then deposited on the face of the brickwork as that water dries. Vigorous work with a stiff brush when the wall is dry will remove this efflorescence. Tests of liability to efflorescence have been formulated, and manufacturers will be able to describe their products in those terms.

Clay bricks are available in a wide range of colours and textures, from dark, low chroma browns and purple-blues, through reds to buffs. Many bricks show a variety of hues within a single brick. Textures are largely a consequence of the particular manufacturing process, and in some cases textures are deliberately induced through a mechanical process. All manufacturers include good colour photographs in the brochures but it would probably be a mistake to specify on that evidence alone. Brick libraries, such as those at the Building Centres, give opportunities to see and compare small panels of different bricks. However, rather like selecting wallpaper, it is best to have seen a larger expanse of a short list of possibles before any final judgement is made.

All clay bricks are made to the standard 215 × 112.5 × 75 mm (9 × 4½ × 3″) format. Some are also made to metric formats such as 200 or 300 × 100 × 75 or 100 mm. (8 or 12 × 4 × 3 or 4″). These all have an unfamiliar proportion and will radically affect scale. Designers who have never seen such bricks in use should be most careful when considering them.

Many clay bricks are available in a range of standard special shapes. Most manufacturers would also be prepared to make special specials. Standard specials are available ex-stock or after a short waiting period. Special specials have to be designed with the manufacturer (largely to ensure that they can, actually, be made) and so there will be inevitable delays in their production. Figure 2.17 illustrates a range of standard specials. Some are designed to facilitate changes in direction of the line of the wall, others to allow variations in thickness, whilst others are for use in particular places such as at the coping or at the end of a wall. Special specials can be very expensive and even standard specials are generally considerably dearer than a matching standard brick. So, whilst special bricks can be most valuable aids to both the practical and visual aspects of a design, care must be taken if budgets are tight.

Clay bricks start to expand from the moment they leave the kiln. Whilst most of this movement will have been completed before they are built into a wall, it can result in buckling if the wall is totally restrained. Expansion and contraction will also occur with temperature changes. Movement joints are recommended at maximum 12 m

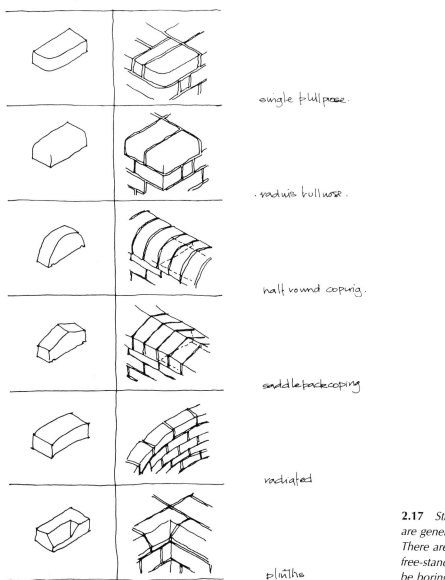

single bullnose.

. radius bullnose .

half round coping.

saddleback coping

radiated

plinths

2.17 *Standard special bricks are generally available ex-stock. There are really no reasons why free-standing brick walls should be boring.*

(40′) intervals. These are most simply formed by breaking the bond so as to give a continuous vertical line of perpends, which are not filled with mortar. If a visible gap is not wanted, then the joint might be filled with a compressible material such as bitumen impregnated fibreboard. But it should be remembered that the movement joint will reduce the wall to being a series of 12 m (40′) long independent walls. Quite apart from considerations of appearance it might be necessary to introduce piers or buttresses at or near movement joints in order that the resolution of one practical problem does not create another.

Damp-proof courses and copings are necessary for all clay brick

161

free-standing walls. Any of the alternatives described in sections 2.1.2 and 2.1.5 would be suitable. Special mention might be made of engineering bricks in this context. Clearly they would be convenient, allowing the same technological approach to the whole wall, and the opportunities to introduce bands of contrasting colour and texture can give class and style to a wall which might, otherwise, have little of either. With slightly higher budgets, the use of plinth headers and stretchers at the dpc can give a psychologically and practically more satisfying base to the wall.

2.2.2 Calcium silicate bricks

Calcium silicate bricks have much in common with clay bricks. This section will, then, concentrate on differences rather than repeat much of the information given immediately above.

Calcium silicate bricks are made by a radically different process than are clay bricks. Whilst the latter are fired in a kiln, the former are steamed under pressure in an autoclave. Controlled gradings of sand and/or crushed flint are mixed with hydrated lime and water. Pigments may be added. This mixture is pressed to the desired shape and size and the resulting units are steamed within a pressure vessel for some eight hours.

Strengths are comparable with clay bricks up to but not including engineering bricks. They are generally formed with frogs (vaguely pyramidal depressions in one bed face) and should be laid frog uppermost. This uses more mortar than if laid frog down, but gives a wall of much greater strength.

Unlike clay bricks, sand and flint–lime bricks tend to contract initially. Moreover, they expand and contract to a greater extent than do clay bricks when subjected to the weather. Movement joints are recommended at maximum 8 m centres (26′). Because of this greater movement, mortars with high proportions of lime are advised; say 1:1:6 below dpc and 1:2:9 elsewhere. (Ratios given as cement:lime:sand.)

Calcium silicate bricks with a crushing strength of 14 N/mm^2 (2 000 lb/in^2) or better, have been found not to be adversely affected by prolonged wetting and drying, or freezing and thawing. They can be damaged by some chemicals which would not affect clay bricks. They can, for example, react relatively badly to atmospheric pollution. But sulphate attack is not a problem provided higher strength bricks are used below dpc.

These bricks have a very precise form with sharp arrises. They can be used with rather thinner joints than can clay bricks, if this is desired for technical or visual reasons. The surface texture of sand–lime bricks is something like that of smooth dressed stone; that of flint–lime bricks is a little coarser. The natural colour is white or off-white, and sometimes cream or very pale pink, depending upon the aggregates used. Pigments can be added – some giving colours that are difficult to associate with bricks!

2.2.3   Concrete bricks

Concrete can be cast to standard brick proportions and then used as described in the general introduction to this section. Various classes are made and bricks for special purposes are needed for free-standing walls. Concrete bricks perform like calcium silicate, rather than clay bricks. The recommendations given for the former in respect of mortar mixes, movement joints, etc. apply to concrete bricks also.

The appearance of a standard concrete brick is less than stunning. Pigments can be added but, as with precast flags, interlocking blocks and *in situ* concrete pavings, the more pastel shades are both weak and timid.

Some manufacturers are producing concrete blocks that are originally cast at twice the width of a standard brick. These are then split in half after the concrete has set to give two standard bricks. This split face has a quality of its own, and has considerable potential. In such cases honest colour can be produced by controlling the aggregates used. Because there are no cultural associations with this product, different proportions to standard bricks seem entirely acceptable. Most manufacturers make a range of lengths and course heights within standard widths. Used single-sized they can give walls of considerable strength-of-purpose. Used to a mixture of sizes they can give a wall that is undoubtedly modern but with strong overtones of tradition – particularly when both the fine and large aggregates in the bricks are crushed stone.

2.2.4   Concrete blocks

Concrete blocks are made, in the main, for use as the inner leaf to cavity brick walls, or for internal partitions within buildings. Their face size is made to co-ordinate with the lengths and course heights of standard bricks. A format size of 450 × 225 mm (18 × 9″) is common. Thicknesses vary with 100, 150 and 200 mm (4, 6 and 8″) being most frequent. They are made to three classes. Class A includes the strongest blocks which can be used in any location. Nothing other than class A blocks should be considered for free-standing walls.

Many of the thicker blocks are hollow. These give a width of wall that is more stable than could be the case with the same weight of concrete in a solid block. These hollows can be filled with concrete as the wall is constructed, leading to great strength. Further, reinforcing rods can be introduced running continuously up inside the wall. This is a particularly valuable detail when a part of a wall needs increased strength but appearance wants to be consistent.

Standard concrete block walls have a distinctly utilitarian appearance, but some are manufactured to have a fair face, and some are cast so as to have both fine and bold textures. Blocks with a range of exposed aggregates are also available. Manufacturers' catalogues should be consulted.

Technical advice for matters such as mortar mixes, movement joint centres, dpc and coping details, are all as for concrete and calcium silicate bricks (section 2.2.1 and 2.2.2).

2.3 *IN SITU* CONCRETE WALLS

2.3.1 Concrete

Concrete as a material has been discussed in section 1.13. It should, then, be sufficient here to do no more than remind readers that concrete has three principal ingredients. **Cement** is manufactured from raw materials (clays and chalk or limestone in the main) allowing the production of calcium silicates and aluminates by blending, burning and grinding to give the familiar pale grey powder. **Aggregates** are all stones reduced (by man or nature) to a range of sizes from fine sands to chippings of perhaps 38 mm (1½″) diameter. **Water** must be free of any impurities and is considered to be suitable if it would also be drinkable. (With some concrete constructions there is a fourth principal ingredient – reinforcing steel.) Graded aggregates include particles with a range of sizes such that the interstices between the larger can be filled by the next size down, and the gaps between those filled with the next smaller, and so on. A batch of well graded aggregate will, then, have many of the qualities of solid rock, the critical difference being that there is nothing to hold the aggregates together. It is the cement, after hydration, which binds the aggregates to form concrete. Exactly what it is that happens when water is added to cement is not fully understood, even by the experts. Microscopic examination of concrete shows that minute systems of cyrstals have been formed which grip the smallest aggregate particles, and because the aggregates are well graded, the smallest grips the next larger, and so on until the concrete has all of the physical qualities of a naturally occurring conglomerate rock. Indeed, its qualities are potentially much better, for the quantities of the various ingredients can be controlled so as to give a concrete that is best suited for a particular use.

2.3.2 Design generally

Wet concrete can be cast into moulds of almost any form. There are four practical limitations. First, hydrating cement produces heat. This can be an advantage when concreting in cold weather but if too large a mass of concrete is cast at one time, the heat produced in the centre of that mass can adversely affect setting. However, masses of such a size are beyond the scope of this book. Secondly, wet concrete is very heavy and has no initial capacity to support itself. The mould will have to provide initial support as well as desired form. If a very complex concrete form was wanted, then the construction of the mould might prove financially or technically prohibitive. Thirdly, it must be possible to remove the mould after the concrete has set. If there was no desire to reuse the mould, it might simply be broken up into small enough pieces. But if reuse was essential (because more identical castings were wanted, and/or so as to get more value from the cost of its construction) then design thought needs to be given to the mould in these terms as well as in terms of the form to be cast. Fourthly, the concrete form must be able to support itself after it has set and the mould has been removed. Concrete can be very strong in compression but does not perform well in tension. That is to say, concrete is difficult to crush but not too difficult to pull apart. Any forces which put a bend in concrete will induce tension stresses. Steel is very good in tension and so can be incorporated within concrete at those places where tension is predicted. It is not our intention in this book to discuss the design of concrete reinforcement in any detail, but designers must be able to recognize when tension stresses might be induced and then modify their design ideas appropriately and/or seek the expert advice and services of a structural engineer.

The sections which follow may seem, at first glance, to start at the end and end at the beginning. (Not an uncommon experience within design.) *In situ* concrete walls have more visual versatility than have any of the other basic forms. By that we mean that two concrete walls could look much less like each other than ever could two brick or stone walls. That versatility is given by the range of finishes available within this technology. But there is an intimate relationship between finish and the nature of the mould. Many finishes are given directly by the mould. Others are facilitated by the mould. With others it is a rather simpler matter of ensuring that the desired finish is not prohibited by the kind of shuttering used. So, an early, broad decision during the design of an *in situ* concrete wall must be about the type of finish that is to be achieved. Only then can the appropriate form for the shuttering be considered and, through that, the ideas about finish developed. Only when a general decision about shuttering has been taken can the more specific constructional details be developed.

2.3.3 Finishes

The finishes which might be given to *in situ* concrete can usefully be considered as falling into two categories of scale. There are those which model the wall as a whole, giving a relatively large-scale pattern or other form of relief, and there are those which give surface colour and/or texture. It is perfectly possible for an *in situ* wall to be finished in both of these ways.

The small-scale finishes which might be given to *in situ* concrete can also be further sub-divided into two technological classes. There are those which are given directly by the nature of the shuttering which might, in part, have been given that nature specifically so as to give that finish; and there are those which are effected by modifying the surface of the concrete after the formwork has been struck. Again it is perfectly possible for an *in situ* wall to be finished in both of these ways, but all must be influenced by at least the first.

So, the overall qualities of finish displayed by an *in situ* concrete wall might be the combined effects of one or both categories of scale together with one or both technological classes.

(i) *Large-scale pattern or relief.* Whilst it is possible to abrade the surface of concrete after it has set and the shuttering has been struck, that would be an enormously time-consuming, expensive and messy operation if significant quantities were to be removed. Indeed, it would be impossible if the concrete were reinforced. Any large-scale patterns or relief must be formed in the shuttering before the concrete is cast. This has several implications for a designer.

As already stated, it must be possible to remove the shuttering after the concrete has set. Technical design of shuttering is not a normal part of a designer's work. Responsibility for adopting the correct construction techniques and practices in this, as with all such matters, is given to (and is best left with) the contractor. However, designers do have a responsibility to ensure that the results of their work will not present the contractor with insuperable problems. Any kind of relief which will result in the set concrete gripping parts of the shuttering is likely to present just such difficulties.

Again, as stated previously, wet concrete is very heavy and the downward and outward loads exerted on the shuttering can be considerable. Consideration might have to be given to constructing a wall in two or more operations, with the later lifts being constructed on top of the earlier, after they have set. But when the wall is to be built in this way, the designer must also give thought to the appearance of the joints between each lift. For practical reasons such a joint would show on a flush wall surface as a ragged line. It is generally more visually satisfactory if the ragged line is disguised by casting a rebate or similar into the face, along that line. Indeed, it could well be that the necessity of disguising the horizontal joint line between two lifts has generated the thoughts about large scale pattern or relief in the first place.

Finally, it must be physically possible for the wet concrete to fill all the available space within the formwork. If it does not, then parts of the intended design will be missing. That would certainly be unsightly and would probably also constitute a technical weakness, certainly so if it left steel reinforcement exposed. There are two reasons why concrete might fail to fill all the intended space. First, if the concrete was expected to find its way through and/or into a space which was smaller than the largest aggregate. Whilst some of the smaller aggregate would find its way into such spaces, the presence of the larger pieces would make it certain that this did not happen consistently. Secondly, if the concrete was expected to defy gravity in order to fill all the available space. Concrete is poured from the top and settles into place from above and the sides. Whilst vibrating the concrete as it is poured will ensure that it is tight against the inside face of the formwork, it cannot be expected to move up into an otherwise inaccessible space – nor can it even be certain that it would completely fill the space below a horizontal length of shuttering.

(ii) Surface texture – given by the formwork. It is strongly advised that section 2.3.2 be read or re-read with that which follows.

In principle, any texture that is on the surface of the formwork will be transferred, in the negative, to the face of the concrete. This principle has to be qualified by saying that the fine aggregates must be able to find their way into that texture otherwise it would be simply expressed in cement which would soon weather away. So, very fine textures cannot be permanently transferred in this way. There might also be difficulties with very coarse textures which might get blocked with intermediate sized aggregates, leading to only partial transfer.

Possibly the best known and most frequently seen example of such transferred texture is that given to board-marked concrete. Particularly grainy lengths of board are used to construct or face the formwork. Sometimes those boards are made even more grainy by wire-brushing or similarly treating them. Both the grain on the boards, and a hint of the butt joint between them, is transferred, as well as, sometimes, impressions of the nail-heads or whatever, used to fix the boards to the shuttering supports. (This technique gives a useful guide to the scale of textures that can be given in this way.) There are other possibilities, but the designer should always check technical feasibility if a novel material or form is being considered for this purpose. The British Cement Association would be happy to answer any queries.

(iii) Surface colour and texture – given by exposing aggregates. The concrete in an *in situ* wall will have gained sufficient strength to support itself in about two days after casting. (That is a rule-of-thumb figure – when a test cube has gained a compressive strength of 2 N/mm^2 (300 lb/in^2) is more scientific.) When the shuttering is first struck the concrete will still contain a great deal of water and its surface (which particularly interests us here) will be quite soft. Moreover, the process of pouring and vibrating the concrete within the formwork will have resulted in a shallow depth of the surface of the wall which contains only fine aggregates and cement. Further, the surface can be encouraged to be even softer than it might have been by using a surface retardant in the release agent which has to be applied to the formwork to ensure that the concrete does not bond to it. So it is that the soft shallow surface can be removed with relative ease after the formwork has been struck. A stiff brush and low-pressure water hose may be all that is needed. Care must be taken for the object of the exercise is to remove only the immediate surface of the concrete, leaving the larger aggregate exposed to view. The precise colour and

textural effects that will be achieved will be determined by the colours and size ranges of the aggregates in the concrete. This was discussed at some length in section 1.13.8 when considering exposing the aggregates in *in situ* concrete slabs. Refer also to the notes in that section about the possible advantages of using gap-graded aggregates.

If the concrete is left to harden, then the aggregates cannot be exposed simply by brushing. Mechanical means must be employed, such as a rotary hammer or grit-blasting apparatus. The clear disadvantages of such methods are that they are expensive and dirty. But there are also advantages. These tools can be used with some precision. Designs can be etched on the surface, leaving some parts of the surface untreated, as a visual background.

Aggregates might also be exposed after the concrete surface has hardened with a hand- or machine-operated percussion hammer. At one extreme, sizeable pieces might be broken from arrisses with rhythmic swings of a hand-held lump-hammer; at the other, the face of the concrete can be given a fine-tooled effect, not dissimilar to that which can be given to stone with gentle use of a small machine.

The combined use of both bold and small-scale texture in an *in situ* concrete wall, together with regular and irregular exposure of the larger aggregates, can make *in situ* concrete an entirely legitimate medium for artistic expression.

2.3.4 Shuttering

The formwork for *in situ* concrete walls can be thought of as a box which will contain the concrete until it hardens. If concrete of good appearance is to be achieved, then it is essential that the formwork does not distort under the weight of the wet concrete. It is also important that the formwork is watertight; for, if not, precious material will be lost from the concrete to harden, in part, as excrescences on the face of the wall. All of this leads to the inevitable conclusion that the formwork must itself be well constructed if the wall is to be so.

The principal materials from which shuttering is made are timber, steel and plastics. The decision as to which is to be used on any particular project is the contractor's – unless specific instructions are given and/or are implied by the nature of the design. But remember that the decision will be based upon the intended finish as well as the number of opportunities there might be to reuse the shuttering. Whilst the decision is the contractor's, the designer must be fully involved prior to that decision being taken.

(i) Timber. Timber formwork is made by the contractor, either on site or in a workshop. It can be manufactured more quickly than can steel or plastics and at a lower cost if it is complex, but it will have a shorter life. For purely utilitarian work, or where a finish is to be given to the concrete after the formwork is struck, a heavy grade

plywood is most commonly used. This can give a firm, even finish, particularly if it has been given a resin-film face which reduces the absorbancy of the ply. Plywood shuttering is particularly useful when the designer wishes to give the wall some individuality, but when a large number of repeats will not be required. Fillets of timber, or other suitable materials, can be relatively easily fixed to the face of the ply by joiners on site – provided that their section is one which can be withdrawn after the concrete has been cast. Figure 2.18 shows a few examples.

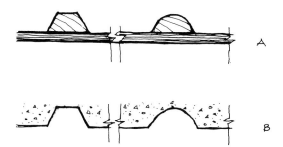

2.18 *A variety of timber fillets (and, indeed, many other materials) can be fixed to the inside face of the shuttering (A) in order to enliven the face of an in situ concrete wall (B).*

(ii) *Steel.* Steel shuttering is most likely to be a part of a formwork system. Whilst simple patterns are available most give a straightforward smooth, even finish. These can be reused many times – indeed would theoretically last for ever if they were not damaged by careless handling. However, the cost of their original manufacture is high, and so it is unlikely that they would be considered for a design with some individuality that was not to have a large number of repeats.

(iii) *Plastics.* Plastic shuttering gives a very high quality of finish with considerable smoothness and consistency. The cost and ease of manufacture falls somewhere between timber and steel. The difficulty with plastic formwork (mainly glass-reinforced polyesters) is that it is more likely to distort under load. So it is most likely to be used when constructing a wall, as a facing to more substantial shuttering.

(iv) *Liners.* Textured plastic or rubber sheets can be used to face the shuttering in order to give that texture to the concrete. These may be purpose-made for a particular project, or designers might find appropriate textures on sheets made for entirely different uses.

Light materials which would have too little substance to support the weight of wet concrete, but with a profile which would be very expensive to give to the shuttering itself, could be used in combination with supporting formwork. The light metal sheets produced to roof industrial buildings, for example, might be used in this way.

Entirely unique patterns and reliefs might be made out of materials such as urethane foams which can be formed, cut and/or melted to give the design in negative. These are likely to be entirely unique, for

169

it could prove impossible to remove them from the cast concrete undamaged. Solvents may have to be used, giving another unique quality, for with this technique textures can be formed from which it would be impossible to remove any other kind of shuttering.

(v) *Permanent shuttering.* Shuttering might be constructed so that it will not only support the concrete whilst it sets, but will also provide the surface finish. Precast concrete panels are a good example of this technique. A consistency of finish can be achieved (using any of the methods described above) because the panels would be made in a workshop or factory. Fixings can be cast into the back of the panels as they are made, allowing the two sides of the shuttering to support each other (aided, perhaps with guylines or relatively few props) whilst the wall is poured; and those fixings also ensure that the face never comes away from its core.

2.3.5 Foundations

Free-standing *in situ* concrete walls should be set on a concrete strip foundation with a breadth and depth as described in section 2.1.1. However, these foundations should differ in one respect. The strip should be formed with a kicker, as illustrated in Figure 2.19. This will give a positive location for the bottom of the shuttering. This is of particular importance if (a) the wall is to have a profiled surface, for the kicker allows the bottom of the shuttering to be straight and flat and, in consequence, better able to contain the wet concrete; and (b) if the wall is to be reinforced, in which case the main bars of the reinforcement in the base are left protruding through the top surface of the kicker and so are available for connection with the steel in the wall, whilst the kicker itself ensures that the reinforcement as a whole is in the correct relationship with the face of the wall. (The formation of kickers is of equal importance, and for the same reasons, when a profiled wall is to be constructed in more than one lift.)

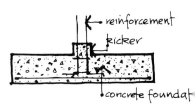

2.19 *The foundation to an* in situ *concrete wall must be formed with a kicker in order to provide a positive fixing for the shuttering to the wall itself.*

2.3.6 Damp-proof courses

There are as good arguments for including a dpc in a free-standing *in situ* concrete wall as in any other. Whilst good, dense concrete is less susceptible to rising damp than are other more porous materials, a considerable thickness of normal concrete is needed before it could be considered entirely waterproof. An unreinforced wall with a generous height to thickness ratio (or other characteristics which ensure stability) might be given a simple first lift from foundation to 150 mm (6″) above finished ground level. One of the dpc materials (described in section 2.1.2) could then be laid on the top surface of that first lift, and the remainder of the wall cast above that. With a reinforced wall this would not be possible or, rather, would be ludicrously time-consuming with holes having to be cut from the dpc strip. In such cases a painted dpc (two coats) can be used, care being taken not to paint the starter bars sticking up from the kicker, for then they would not obtain any bond with the concrete. A simpler alternative might be to use a waterproof additive in the first lift of concrete from found to dpc.

2.3.7 Copings

The top of an *in situ* concrete wall has to be kept dry for the same reasons as must any other. Any of the copings discussed in section 2.1.5 could be considered, with due regard to visual compatibility. Alternatively, a coping could be formed *in situ* with its form integrated with the relief on the wall below.

Plain *in situ* concrete does not weather terribly well. The brightness which is in the newly cast concrete soon dulls, and the material is particularly prone to streaking where rain-water runs off down the face. The provision of a coping with an effective drip is, then, most important, but this is rather less of a problem with a modelled and/or textured wall where the course of the run-off is guided by that texture, and the streaking can amplify pattern rather than disfigure the surface.

2.3.8 Expansion joints

An unreinforced *in situ* concrete wall would need to be given expansion jonts at 4.5 m (15′0″) centres. The wall is cast as if the expansion joints were to be the end of the wall, and that end face covered with a 13 mm (½″) thick strip of bitumen-impregnated softboard or similar proprietary material, before the next length is poured. Alternatively, a rather larger gap might be formed and left unfilled. A reinforced wall needs expansion joints at 9.0 m (30′0″) centres because the reinforcement will help contain movement due to initial contraction and, later, temperature changes.

Remember that it must be possible to connect the next run of shuttering to the previously cast length of wall in a way which is both sound and waterproof. The vertical equivalent of a kicker will be needed and any large- or small-scale texture must stop short of the ultimate expansion joint, by 100 mm (4″) or thereabouts. A good designer would incorporate this necessary feature within the design of the wall as a whole.

2.3.9 Initial contraction

All concrete contracts as it hardens. This sets up stresses within the material which can lead to cracking. Whilst this would not be structurally damaging (given that contraction joints have been included within the construction) it would disfigure any intended surface pattern. Reinforced concrete does not suffer this problem, for the steel distributes the stresses evenly throughout the wall. For this reason it is well worth while including a sheet of light mesh reinforcement in all free-standing, *in situ* concrete walls, positioned about 25 mm (1") back from each face, even when they need no additional structural strength.

2.4 RENDERED WALLS

Rendered free-standing walls are less common than with buildings proper. This is because rendering is rather more vulnerable when used on a free-standing wall. However, given that it is properly executed and maintained, there are a number of instances where it provides a useful answer. And, given that the technology is understood and the wall well constructed, there is no reason why the results should not be perfectly satisfactory.

The integration of internal and external spaces by giving them related forms and defining them with similar elements can be a valuable strategy. If the buildings associated with those external spaces are rendered then the use of the same finish on the free-standing walls might well be considered. Where budgets are low but ambitions high, rendered walls are one of the few available options. The wall itself can be constructed with common bricks or concrete block (and no wall could be constructed cheaper) and then finished with a material with genuine technical and aesthetic qualities. Moreover, this generates no philosophical difficulties – the approach is not dishonest, for this is precisely the way in which a rendered wall wants to be built.

If a painted finish is wanted for the wall (for a mural perhaps, or in order to improve qualities of light through reflection) again render is one of the few available options. The only alternative surface that is really smooth would be *in situ* concrete; but that has a surface density that needs special treatment if it is to take paint. Textured renders can be painted to equal effect. Painted brick and block walls can look very well, but inevitably the joint lines show through. Whilst it is true that very light colours will need regular repainting if their freshness is to be maintained, painted rendered walls should not be left off the designer's palette. (But note that special types of paint are needed – refer to manufacturers' catalogues.)

Last, but not least, rendered surfaces can offer a range of textures and patterns of kinds which cannot be achieved through the use of any other technique. Rendered free-standing walls have, then, a validity in their own right.

2.4.1 Render

(i) Materials and mixes. The material used for rendering has a basic similarity with the mortar used in masonry walls (section 2.1.4).

Proportions of cement and/or lime with fine aggregates are mixed with water and then spread over the wall in thin layers with a steel or wooden float. (A float is like a bricklayer's trowel but rectangular rather than triangular in shape.) The actual proportions used should vary to suit the background material that is to support the rendering. The more likely it is that the background will move with temperature variations, the softer must be the render mix. If ever in doubt, it is better to specify a softer rather than harder mix. Strong mixes (those high in cement) will shrink as they harden and this is very likely to lead to cracking. An uncracked render will remain waterproof even if it has an essential porosity, whilst a dense render will have no such qualities when it has cracked.

(ii) *Backgrounds.* Most walls can be rendered, but bricks containing sulphates may cause trouble if they get damp. It is important that there is a key between the first coat of render and its background. Without such, there would be no bond between the rendering and the wall, and it would fall off. Porous backgrounds may well provide sufficient of a key without further attention. Denser backgrounds may have to be given a mechanical key. As examples, an *in situ* wall made of concrete with little if any fine aggregate will provide a perfectly adequate background; lightweight concrete blocks and soft bricks will be satisfactory in themselves, but it is best to rake back the mortar in the wall's joints to a depth of 10 to 13 mm (⅜ – ½"); with denser bricks it is essential that the joints are raked; whilst with very dense bricks or an *in situ* wall made of normal concrete, a mechanical key of splatterdash would need to be given as a preliminary treatment. Splatterdash is a very strong mortar mix with 1 part of cement to 2 parts of sand, mixed very wet. Proprietary additives to improve its adhesion might be included. This is dashed on to the wall (often by vigorous flicks with a large soft brush) to give a rough and deliberately uneven texture.

(iii) *Rendering.* The wall should be well brushed down and thoroughly damped to ensure that suction will be even. It is important to keep the background damp (but not running wet) otherwise the water in the mix will be sucked out into the wall behind, leaving too little for the chemical reaction in the render. Newly built walls should be given an opportunity to dry out (minimizing the amount of water trapped deep in the wall and ensuring that any efflorescence will be brushed off when the background is prepared). A 1:3 to 1:1:6 undercoat (depending upon the nature of the wall) is spread on the background to a depth of about 15 mm (⅝"). This must be scratched or combed, in order to provide a key for the next coat, before it starts to dry. The undercoat should be left for at least 24 hours so that it has achieved its initial set and shrinkage. The render might then be finished with one more coat if it is to be smooth, or two more if it is to be textured.

173

A smooth finishing coat would use a mix that is certainly no stronger than that of the first (and, perhaps, a little softer) spread to a depth of up to 10 mm (⅜″) and floated smooth as the work progresses. If the render is to be textured, the second coat would also have to be scratched or combed to provide a key.

A rough-cast texture (also known as harling) uses a mix of 2 parts of cement to 1 part of lime to 2 parts of fine aggregates. In this case those fine aggregates might be up to 13 mm (½″) in size, but graded to that size from sand. The larger the stone chippings or whatever, the rougher the texture. Prior experimentation is well worth the time and trouble involved if a pre-existing sample is not available for inspection.

A pebble-dash or dry-dash finish can give both colour and texture. The second coat is put on a little thicker than for a smooth finish but using the same mix. Washed and drained spar chippings or shingle of up to 10 mm (⅜″) are dashed on to the surface as work proceeds.

A number of proprietary mixes are available for the finishing coat. These should always be applied as set down in the manufacturers' instructions. Those instructions are likely to include recommendations for undercoat and preparatory treatments for different backgrounds. Some of these proprietary mixes are machine applied. The simplest of such machines looks rather like a hand-cranked hair-dryer. The mix is placed in the body of the machine and, as the handle is cranked a comb of springy steel flicks small quantities on to the rendered wall. Such a machine can be used for non-proprietary mixes such as described for rough-cast, although a maximum aggregate size of 6 mm (¼″) would be the limit in this case.

If the smooth rendered surface is left for a few hours so that it has started to set but remains malleable, then it can be given a host of different patterns/textures. These all involve either scratching with a plasterers' comb, old saw blade, or similar, or imprinting the render with any manner of objects or tools. Discussion with the craftsmen and discovering their experiences can often be more valuable than a more dictatorial specification.

(iv)   Colouring. Additive pigments can be used with render but are hardly necessary. Because the material contains a high proportion of fine material, the colour of those fines will affect the colour of the render as a whole. Experimentation is advised. If darker or high chroma colours are wanted, then the use of a proprietary mix is likely to be most successful.

2.4.2   Rendered wall details

The foundations for a wall that is to be rendered should be exactly the same as they would have been otherwise for a wall of that background material.

A dpc will be needed and it is advisable not to take the render below that dpc. If the dpc were rendered over, then the render would certainly crack at that line. And, if the render were to be taken close to the ground, it is very likely to get wetter than would be preferred. However, it may not be desirable to be able to see the background wall below dpc. A common response to this difficulty is to have the wall below dpc constructed fair faced and to give it two coats of bitumastic paint. This can leave the rendering visually floating above the ground – to some effect. Alternatively, a different material such as facing brick or stone, might be used for the courses between dpc and just below ground level.

Finishing the render at the dpc must be considered. A batten can be temporarily fixed to the wall below dpc to give a positive edge to which to render, and the rendering itself finished with a bell-mould to act as a drip and to give a slight visual emphasis at this point (which stops the wall looking top-heavy). Commercially available light metal beadings with expanded metal strips attached facilitate this kind of detail. When made of non-rusting material, these are more expensive but much more durable (Figure 2.20).

The background wall will have to be built with any necessary expansion joints exactly as it would have been had it not to be rendered. Again, it is important that the render is not carried over those movement joints. The kind of expanded metal beading described above has a value here also. Filling the gap between two render stop-beads at a movement joint should be seriously considered. If water in any quantity should start to penetrate behind rendering, it will fall off at the first frost. The channel formed between two stop-beads would almost guarantee that this would happen. Figure 2.21 shows how that channel might be filled with a gunned non-hardening mastic, which will provide an effective seal.

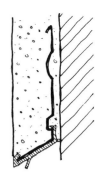

2.20 *Proprietary items are available to contain the perimeter of areas of rendering. The example shown here forms a drip at the bottom of the rendering.*

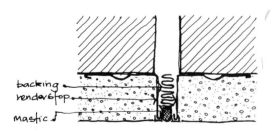

2.21 *Rendering would be certain to crack if it was carried over expansion joints in the wall behind. The detail shown here uses pressed metal items made specifically for this purpose.*

It is equally important that water does not penetrate behind the rendering at the top of the wall. The use of a correctly designed coping is possibly more critical with this finish than with any other. If the coping is to be sat on a dpc, then the kind of detail described for movement joints might be considered. Alternatively, a coping which has rebates, in addition to those which act as a drip, which will accept the top edge of the render might be considered. But by far and away the most effective technical detail would be to use a pressed metal coping designed to cover the whole of the top of the wall, background and render together. Figure 2.22 illustrates alternative copings to rendered walls.

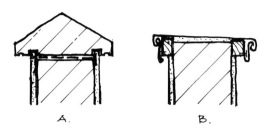

A.   B.

2.22 *It is of critical importance that water cannot penetrate behind rendering. If it does and then freezes, the rendering will fall off. This can be a particular problem at copings. Stone or precast concrete copings can be rebated to contain the render thickness (A) whilst pressed metal copings can form a drip well below the top of the render (B).*

2.5 STONE WALLS

Building stones can exhibit a wide range of characteristics. Some are so soft that they can virtually be scratched with a finger nail, whilst others are so hard as to stand for centuries without any apparent change. Some dissolve over time through the action of rain, others spall to reveal contours or become ridged through the action of the weather. Some are so light coloured as to be reasonably called white and others are as near black as ever occurs in nature. Some have a grainy texture and others a crystalline structure. Some can be given a high degree of polish whilst others would simply rub away if any such attempt were made. Clearly it is not possible, in this book, to describe all of these characteristics for the stones from each working quarry. Designers should make reference to the stone they see in use in a particular locality, to the contractors they find using those stones, and to the quarry masters from where those stones have been obtained. They should then design within the basic principles which are outlined here.

2.5.1 Types of stone

Building stones can be divided into three broad classes – igneous, sedimentary and metamorphic.

Igneous rocks are all called granites in the building trade (although many would not be described as such by a geologist). They are formed when molten magma (the material on which our continents float) reaches the surface of the earth and cools. All are basically crystalline with the size of those crystals varying with the rate at which the magma cooled; the more rapid the cooling, the finer the grain. Colours vary from black and deep red, through greys and greens to dull white. Many sparkle due to the presence of quartz and mica. Others have a quasi-sparkle given by deep black crystals of hornblende and augite. All are immensely strong and durable but, in consequence, are very costly to quarry and work. All can be polished to a mirror-like surface.

Sedimentary rocks give building sandstones and limestones.

Sandstones are tiny particles of other rocks which have been broken down by the action of the weather, then carried by rivers and streams, to be redeposited in lakes and the sea. (Some very fine particles are carried by the wind to be deposited in drifts.) As the particles come from a wide range of sources, the chemical composition of sandstones can vary considerably. The particles are cemented together by minerals originally held in solution in the lake or sea, and consolidated by the weight of superimposed deposits.

Limestones were similarly deposited in lakes and seas but are built up from the remains of small marine and aquatic animals. Their skeletal origins have given them a high calcium content.

Whilst areas of sedimentary rocks will have been deformed by subsequent earth movements, the original bedding planes remain within them (although no longer, necessarily, horizontal). The stones are relatively weak along these planes, allowing their convenient reduction to buildable sizes. The intervals between the bedding planes can vary considerably. Some are far too close to give a stone of useful thickness if it were split on them all. Care must be taken when using such stone to ensure that it is laid in the wall on its natural bed. If not, there is a real risk that the stone will delaminate due to the action of the weather, or under load. Other stones (generally referred to as freestones) have their bedding planes much further apart. These are invaluable when large blocks are wanted for use where they will be exposed to the weather on all sides.

The colours of sedimentary rocks are mostly in the light to dark buff range, although this can be near white as with Portland limestones, or orange-red when oxides have stained the rock. All darken significantly as they weather. None have the kind of strength found in a granite. All can be dressed to a flat, even surface but none will take a polish.

The textures of sedimentary rocks can vary considerably. Many have a very fine granular appearance whilst others more clearly exhibit their origins. Some have pebbles embedded within them. Generally speaking, limestones have a finer texture than do sandstones but the oolitic limestones such as found in the Cotswolds have a fish-roe quality resulting from being deposited in a tidal area. Many sedimentary rocks contain fossils which can be an interesting decorative feature.

Metamorphic rocks consist of older rocks which have experienced structural change by being subjected to great heat and pressure either due to earth movement or because of adjacent igneous activity. There are three groups of metamorphic rocks which are useful as building stones – slates, marbles, and quartzites.

Slate is clay metamorphosed through pressure and associated heat. As the slate cooled again it formed planes of crystals in line with the pressure. These are quite distinct from any bedding planes within the clays. The slate splits very easily on these cleavage planes. This makes it very valuable as a building material but there can be a danger of further delamination through differential thermal movement between a sun-exposed surface and the cooler material below, and subsequent action of frost. Slate can be sawn to a fine, smooth surface and can be given a dull shine. Most slates are a deep blue-black colour although some have distinct hints of purple and others are a quite bright green. In all cases apparent colour is often lighter than true colour, due to surface reflectivity.

Marbles are metamorphosed limestones. True marbles solidified from a virtual liquid and are a fine crystalline white with veins and markings given by other minerals. Other stones (still called marbles in the trade) were less completely metamorphosed and are duller in appearance. All are strong in compression but may split on the veins under tension. (When this happens during construction and that piece is valuable in itself or because its figuring is to be matched with adjacent pieces, it is often simply glued back together using one of the many modern synthetic adhesives.) The colour variation between the parts is as wide as in any paint manufacturer's catalogue. Marbles must be inspected and selected in the quarry or wholesaler's yard – nobody would deliver stone specified simply as marble. All can take a polish but will lose it again if the atmosphere is polluted. They are all expensive (some extremely so) and so are most frequently used as thin facings to some other structural material.

Quartzites are metamorphosed sandstones. They are relatively uncommon. They are even harder and more durable than granites but with cleavage planes and, sometimes, vestigial bedding planes making them easier to work. The riven surfaces are often uneven but can be polished. Colours vary through greys, greens and golds, with a lustre rather than sparkle. Cost again leads to their more normal use as floor slabs or wall tiles.

2.5.2 Forms of building stone

The forms in which building stones are available vary over a spectrum from as quarried at one end to very precisely dressed slabs at the other. (But not all types of stone are available in forms across the whole of this spectrum.) For convenience, this spectrum is divided into two parts, with each relating to one of the forms of construction described in section 2.5.3.

   (i) Rubble. Stone **as quarried** or indeed, as found, is most frequently described as rubble, although there are some local variations in terminology. The form which stone rubble takes varies with the type of rock from which it derives. Sedimentary rocks and those metamorphic rocks with cleavage planes, will give a rubble with two parallel faces where the rock has broken on the bedding or cleavage planes. Clearly such stones have an immediate building potential. Such rubble may be very roughly dressed in order to knock off any obviously inconvenient projections and to give a generally rectilinear

face. The material is then described as being squared or roughed rubble. Granites and marbles do not break in this way and the resulting irregular forms are of little immediate value. Nor can they be easily rough dressed. However, those which have been quarried by the weather become rounded as they are tumbled downhill by storm water and moved by tides and waves. These eroded forms (cobbles) do have a value for rubble walling.

(ii) *Ashlar.* All stones can be **dressed**, with the method employed varying with the hardness of the rock. Softer stones can be sawn, as can those which are harder but which naturally split into relatively thin slabs. Originally sawing was done by hand but it is now almost exclusively a machine operation. Harder stones were originally dressed with hand-held and then mechanical percussion hammers but modern equipment now also allows them to be sawn. To this should be added that very valuable stones have been very carefully treated and laboriously sawn for centuries, irrespective of their hardness. The resulting blocks of stone are most frequently described as ashlar.

Traditionally there was little or no dimensional co-ordination between the ashlar blocks from different quarries, but normal quarrying practice was to cut the stone to a range of course heights as determined by the nature of the rock being worked. When required for specific projects, blocks of identical size were, and are, cut – but at a price. Nowadays there is a tendency for building stones to be cut to a standardized range of sizes. This is not so much a result of a conscious decision taken by quarry owners collectively, but of each responding individually to a common influence. Increasingly the building trade is using stone in association with concrete blocks when building walls. Many quarries now cut their stone to course heights which, in combinations, co-ordinate with the course heights of standard concrete blocks. One, for example, offers ashlar sawn to three course heights – 67 mm, 143 mm, and 219 mm (2⅝, 5⅝ and 8⅝"). Three courses of the first of these (when an allowance is made for the thickness of the mortar beds) are equal to the last; one course of the second with one course of the first, is also equal to the last; and the last is equal to the course height of concrete blocks. Lengths generally, unless to special order, are random and are as the material allows. The manufacturer quoted above, for example, who is selling an oolitic limestone, offers random lengths between 125 and 350 mm (5–14").

The face of the ashlar blocks can be dressed in a number of different ways. That which is closest to a rubble is described as split, or perhaps split and pitched if the arrisses between the split and sawn faces have been knocked off. A split face can be hand- or machine-worked to give a smooth chiselled margin. The commonest all-over texture is fine sawn, which is that given by the basic process. This might be further worked in a number of ways to give rilled or punched faces, for example, which again might also have a chisel dressed margin. (It should be added that these finishes are effectively limited to the softer stones. The cost of treating hard rock similarly would be astronomic.)

Very few if any modern buildings are made of solid stone. New buildings which might, at first glance, appear to be built of stone are almost certainly no more than faced with ashlar slabs held back to a reinforced concrete (or some other) structure with metal ties. Indeed, there is a long tradition behind this approach, using more expensive carved and dressed stone to face a structure of much cheaper stone or brick.

The sizes to which ashlar facings are cut vary considerably with rock type, but there is a tendency for ashlar facings to be rather larger than ashlar blocks. Thicknesses might be as little as 25 mm (1″) with slates, to 75 mm (3″) or so with softer rocks.

Because they are non-structural, it is possible to use ashlar facings in ways which would not, otherwise, be recommended; with bedding planes which are not horizontal, for example. But it must be remembered that they will still have to withstand the weather. Whilst deterioration due to differential expansion and contraction, through the action of frost, or resulting from the migration of salts after repeated wetting and drying, for example, might not produce practical difficulties, they would certainly lead to visual disfigurement.

2.5.3   The construction of stone walls

(i)   *Rubble walls.* Random rubble walls use the material directly as quarried or as found. They are almost always laid dry – that is without mortar of any kind. Dry stone walls are a superb example of a construction which originally had no aesthetic pretensions whatsoever but which is now seen to have great beauty. This beauty comes from a straightforward use of a simple technology which inevitably responds to the landscape of which it is a part, for the materials always were a part of that landscape.

Details vary from county to county as traditions evolved which put to best use the material to hand. Most have battered faces (are thicker at the base than at the coping) to offset the effect of gravity. Many are constructed with two skins with the occasional through stone, with the core of the wall packed with smaller and finer material. Others, where larger slabs are available, have more through stones and may be exclusively through stones. They are then often laid so that the individual stones all slope slightly to one side, in order to throw off water. Figure 2.23 shows a typical section.

Copings also vary with location and tradition. Where the material allows, a simple capping slab may be used. Where the stones are smaller, an approach known variously as 'lords and ladies', 'soldiers and sailors' or 'combers' (pronounced, in Gloucestershire at least, with a short 'o' and silent 'b') might be used. With this, roughly triangular or trapezoidal stones are stood on end, vertically or at a

180

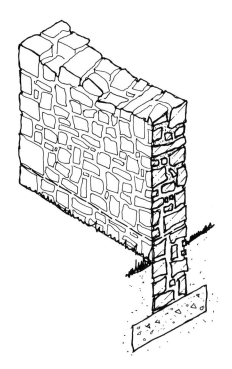

2.23 *Random rubble stone walls have no obvious coursing. Note the larger stones, or quoins, shown in this example at the end of the wall. Note also the occasional through stones which tie the two faces together. Finally, note that this wall is unfinished – the coping is yet to be built.*

slight angle, on top of the body of the wall. Whilst this could never be satisfactory for other kinds of walls, it is sufficient to throw off a heavy rain storm and so avoid the core of the wall being disturbed.

Dry stone walls never need a constructed foundation, for it will be naturally provided just a short distance below ground. We would not encourage the building of dry stone walls anywhere where this was not the case, not because that would be technically impossible but because it would be aesthetically absurd. The nature and location of their construction means that dry stone walls are not given a dpc.

Random rubble walls have no apparent coursing. If the material can be easily squared or roughed, then courses start to emerge. If the stones are of consistent thickness, then a coursed rubble wall can be built; if of inconsistent thickness, then the wall can be brought to courses at intervals up the wall. Figure 2.24 illustrates this alternative.

Coursed rubble walls may also be dry when edging a field. But as their appearance becomes more sophisticated, they are more likely to move into the villages and towns, and then they are also more likely to be built with mortar joints. As with other masonry walls, the closer the performance of the mortar can be to that of the stone, the more satisfactory will be the result. For soft limestones and sandstones, a mix of 1:1:6 (cement:lime:sand) would be adequate, and 1:2:9 might be even better if building well before frosts might be expected. The harder stones might need a strong 1:3 mortar, without lime.

Traditionally, a coursed and mortared squared rubble wall would undoubtedly have been given a dressed stone coping as described in section 2.1.5 where overall quality was to be expressed. Others might simply have been given a rounded or weathered mortar cap. Yet

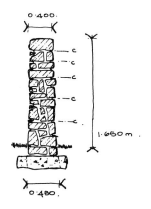

2.24 *Rubble walls can be brought to courses at intervals which frequently coincide with the height of the quoins. With the best traditional examples coursing intervals (C) diminish with height.*

others would have used combers but with mortar to both bed and perpends. All these options are still open, and the choice can be augmented through the sympathetic use of non-traditional details. For example, the crisp line of a pressed metal coping would contrast well with the rugged qualities of the stone, whilst leaving the visual emphasis on the wall itself.

The foundation might be no more than an initial course of larger slabs set down at an appropriate level in order to give a consistent base on which to build or, if such stones are not available, a concrete strip designed as described in section 2.1.1. Any dpc of the kinds considered in section 2.1.2 should be included, although slate or engineering brick might look a little odd unless that oddity was reinforced to become a feature, by repetition at the coping. If fully restrained, movement joints should be included in a mortared stone wall, at 8–10 m (26–33 ft) centres.

Harder stones with cleavage planes can be used in a mortared rubble wall which uses those cleavage planes as the face. The irregular outline of each stone gives a distinctly different character to those described above. The scale is much larger than with other walls of comparable height, and the visual experience is more staccato. A strongish mortar is needed, not only because the stone is hard but also because there is little structural stability in the way in which the stones are used. The inclusion of some through stones would help considerably in that respect. The face of such a wall is so active that it always appears more satisfactory when it has only a thin coping, or no distinguishable coping at all. Foundations and dpc are needed as before.

(ii) Ashlar block walling. There is, in fact, very little left to be said about ashlar block walling specifically. Technically it has a great deal in common with rough dressed and mortared rubble walling. However, the regular dressing does allow the individual blocks to be either more, or less, emphasized within the wall as a whole. Being cut to size, there will be much less dimensional variation between blocks than, say, is the case with clay bricks. Mortar joints can be much thinner, for they do not have to accommodate any such irregularities. Thin and carefully pointed joints, when used with a fine sawn or rubbed faced stone, throws all the emphasis on the plane of the wall as a whole; with the joint lines being no more than a geometric pattern traced faintly over that face. On the other hand, and by way of example, if the ashlar is given a heavily punched face within a meticulously drawn chisel margin, then emphasis returns to the individual block. The use of such blocks as quoins, piers and buttresses, whilst the body of the wall is fine dressed, with the whole being founded on a strong plinth and contained above with a generous coping, can give a wall immense dignified self-assurance.

Technical matters, such as foundations, dpc, copings and expansion joints, are all as described in sections 2.1.1 to 2.1.8.

(iii) Ashlar facing slabs. Thin ashlar blocks can be used to face a stone wall with a thickness which would be unnecessarily expensive to build totally in ashlar. The core of the wall should behave in the same ways as will the face – and in many historical examples this is simply a rubble wall of the same stone as the ashlar. The principal concern is that the face should not peel away from the core. This is done by bringing the rubble core to courses at each course height of the ashlar face. Butterfly wire or fishtailed strip ties are then built into the mortar joints of both the ashlar and rubble, at about 900 mm (3'0") intervals along the length of the wall. These ties should be non-ferrous, stainless steel, or plastic – otherwise on rusting they will expand and split the stone, and disfigure its face with orange-red streaks. Technically, such a wall should be designed within the constraints applicable to all mortared stone walls.

When used to face a structure other than masonry, ashlar stone facings are usually held away from that structure by cramps which are cast or otherwise built into it. This allows the structure to be built and then faced as a separate operation later, when there is less risk that the ashlar would be damaged or stained. It also allows the ashlar and the structure to move differentially with little danger that they will damage each other in the process. There are many proprietary fixings on the market designed for use in just this circumstance.

It is somewhat unlikely that a free-standing wall would be built using this technique, unless it was that it had to exactly match an adjacent building. The details for the free-standing wall should then be developed in concert with those for the building.

Technical details such as foundations, expansion joints, etc. should be those which are appropriate to the structure, provided that the ashlar face has been allowed to move independently.

(iv) Stone banks and hedges. Figure 2.25 shows a typical, traditional stone bank. These, when first built, have many of the qualities of a dry-stone wall (albeit rather more bulk) but the stones rapidly disappear behind a cloak of grasses and wild flowers. The first edition

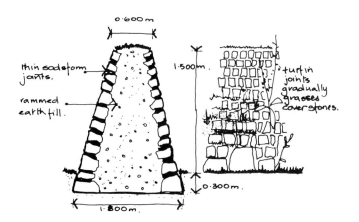

2.25 *Stone hedges are built with two faces of battered rubble stonework bedded on thin grass sods, with the centre filled with rammed earth. With time a stone hedge disappears behind naturalizing vegetation.*

of this book pleaded for their reintroduction in rural areas where new isolated buildings needed to be effectively rooted into the broader landscape. Their value in that respect remains as high as ever and, further, the increasing appreciation that the urban environment offers as many (sometimes more) opportunities to provide wild life habitats, should argue for their use in many contexts where space allows.

Figure 2.26 shows a close cousin of the stone bank, a turf hedge. The substitution of turves for stones limits height, which is then frequently made up with a post and wire fence. This detail, perhaps, would be of rather less value in an urban context as it is very casual in appearance and particularly vulnerable to damage if climbed or bumped.

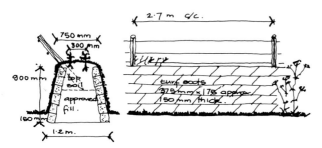

2.26 *The construction of a hedge bank follows the same principle shown in Fig. 2.25 but using turf instead of stone to form the faces. This limits height and so a simple post and wire fence is often incorporated when stock has to be contained.*

2.6   TIMBER FENCING

Timber fencing has the advantage of robustness, durability and straightforward maintenance. Repairs are simple and require materials which are usually readily available. Moreover, timber allows the design of fences with radically different characteristics. It is a fallacy to assume that because a natural material is being used, it must inevitably display a rural quality. It would be as wrong to assume that any timber fence would be suitable in the country, as it would be to assume that no timber fence could be satisfactory in a sophisticated urban context. Everything depends upon how the timber is used. Straightforward detailing and sound construction will always give a fence which looks well in itself; and the material has sufficient design flexibility to give a fence which both looks well in itself and well in its context.

The life of wooden fencing, exposed to all kinds of weather, will depend upon the natural durability of the timber, the way in which the material has been treated with preservative, its technical design, the skill of the people who erect it, how it is finished, and how it is maintained.

2.6.1   Timber

   (i)   Types. The building trade divides timber into two main classes;

hardwoods and softwoods. Hardwoods come from deciduous species (although not all give usable timber) and softwoods come from coniferous trees. As with stone, there can be some difficulties with nomenclature. The names commonly used would not always be seen as strictly accurate by a botanist. As examples, Douglas Fir is not a fir but of the genus Pseudosuga; Western Red Cedar is not a cedar but of the genus Thuja; and Parana Pine is not a pine but of the genus Araucaria. Further, redwood, red deal, yellow deal, red softwood, yellow softwood, Norway Fir, and Scots Fir are all names variously given to timber from the species Pinus sylvestris or Scots Pine. However, it is not often the case that softwoods are specified by species and so this potential confusion can generally be avoided. There are few such confusions with hardwoods which are generally given the name commonly used in the places where they are grown.

There can be considerable differences between timber from different species, between timber from the same species but grown in different parts of the world, and between timber from the same individual but from different parts of the tree. These differences affect the structural qualities of the timber and its appearance. Appearance with hardwoods is often a critical factor in their selection whilst with softwoods it is more often their practical qualities (or degree of freedom from defects) which is most important. Defects can be in the timber itself or can be a consequence of inept conversion (cutting into planks) or seasoning.

The growing tree might have suffered mechanical damage – from fungi, insects, other animals, the weather – resulting in scar tissue. The tree might have been growing on a steep hillside, giving brittle tension wood on the uphill part of the lower trunk. There might have been a breakdown of the cell walls within the trunk, given the various shakes illustrated in Figure 2.27. Pockets of resin may have accumulated, giving timber which is difficult to work and impossible to paint. And, the timber will inevitably have knots where the branches were rooted in the trunk. Whilst these are not defects in themselves, they would be if the branch had died (leaving the knot loose within the timber) and/or if the knot appears at awkward places on the converted timber.

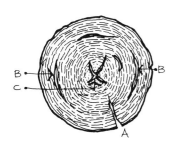

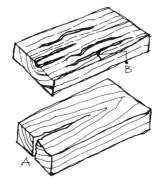

2.27 *The trunks of trees, as they grow, can split in a number of ways, These give shakes; this figure shows radial (A), ring or cup (B) and star (C) shakes.*

(ii) Conversion. Conversion is the basic sawing of the timber into planks. Figure 2.28 illustrates five common methods of doing this. The more expensive the timber (meaning the hardwoods rather than softwoods) the more complex its likely conversion. There are two objectives. First, to minimize waste and secondly, to maximize the amount of timber cut radial to the trunk. Radially cut timber will be stronger in use and less likely to distort.

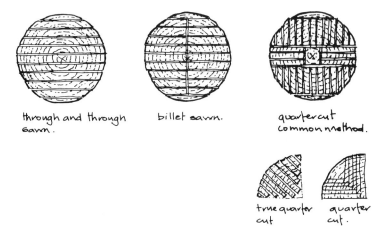

through and through sawn.

billet sawn.

quarter cut
common method.

true quarter cut

quarter cut.

2.28 *Timber can be converted (cut into planks) in a number of ways. The more valuable the timber, the more careful its conversion.*

Inept conversions can lead to one or both of two types of defect. If timber is cut too close to the growing outer surface of the trunk the resulting boards will have a waney edge. The sharp arris seen elsewhere becomes rounded and the brown, soft material of wane can present both visual and practical problems. Figure 2.29 shows a waney edge. No tree trunk is cylindrical. All taper from their base towards the top of the tree. It is, then, inevitable that the rings of cells by which the trunk diameter increases annually, lead to a sloping grain when that timber is converted. Timber where the grain is parallel to the edge of the board will be much stronger than when that is not the case. A grain angle of 1:10 is generally acceptable but 1:20 might be specified where the timber is to be highly stressed in use. Inept conversion can lead to defective boards with unacceptable degrees of slope in their grain. Further, such boards are more likely to distort during seasoning or in use.

waney edge

2.29 *Planks which are cut too close to the edge of the trunk will show a waney edge.*

(iii) Seasoning. Seasoning is the controlled reduction of the moisture content in the converted timber. Ideally, the moisture content of the timber should be that of the environment in which it is to be used. For internal work a moisture content of 10–12% is required whilst for external work 17–23% is satisfactory. (Note: the vast majority of timber offered for sale has been seasoned assuming an internal use. Timber which is too dry for its location can be as much of a problem as that which is too wet.) Timber expands and contracts significantly with changes in its moisture content. The resulting movement can

186

mean that joints open up on contraction, or timbers warp or bow on expansion.

Two methods are used to season timber. With air seasoning, the boards are stacked in a way which allows free air movement through the stack, which is then protected from rain, direct sunshine and ground water without inhibiting the wind. The moisture content of softwoods will be reduced to 17–23% in 8 to 10 weeks in the most favourable conditions. For hardwoods this period is more likely to be a year to eighteen months. Clearly, air seasoning would be most satisfactory for landscape work. It is not possible to reduce moisture content in timber to that experienced within a modern centrally heated building by air seasoning. The timber must be kiln seasoned. The timber is stacked as before but is fully protected from the elements, and a warmed and de–humidified wind is artificially induced.

The possible defects illustrated in Figure 2.30 are all more likely to happen if seasoning is rushed. All result from stresses set up within the timber due to unequal rates of drying. Whether it is cupping, twisting, springing and/or bowing that occurs, it is dependent upon the grain pattern in that particular piece. In addition, seasoned timber can become case hardened if kilned too quickly or if exposed to direct sunlight (giving inconsistent working qualities through the section),

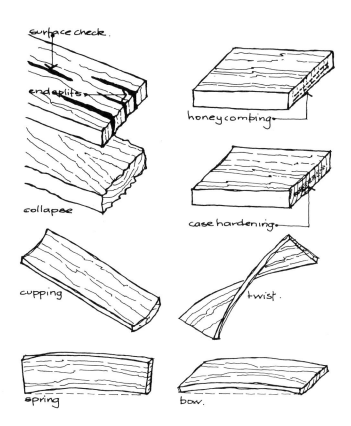

2.30 *Timber can develop defects through seasoning, particularly when this is rushed.*

187

and air seasoned timber can suffer various defects to the end grain exposed at the ends of the stack. This last is easily rectified simply by sawing off the defective material.

It is important that designers understand the range of possible defects that timber might exhibit. In some circumstances they may not be defects at all but qualities which are of no consequence or, even, ones to be admired. Timber should be specified solely in terms of those defects which it is critical that it should not display – free of loose knots, for example – when other possible defects are of no consequence; and other qualities mentioned only when they are specifically wanted – waney edged boards, for example.

(iv) Knots. Timber with large frequent knots would be unsatisfactory for structural work but might be sought after for its decorative potential. The positions of the knots on the boards' faces and arrises further affect potential strength. A sound knot would present few problems but if at all unsound and/or located where it would be subjected to tension, could lead to difficulties.

(v) Sizes. Softwoods are converted to a range of standard sized sections. The more common of these, with a particular value in landscape work, are shown in the table below.

Thickness mm (in)	Width mm (in)								
	75 (3)	100 (4)	125 (5)	150 (6)	175 (7)	200 (8)	225 (9)	250 (10)	300 (12)
25 (1)	x	x	x	x					
32 (1¼)	x	x	x	x					
38 (1½)	x	x	x	x	x	x			
50 (2)	x	x	x	x	x	x	x	x	x
63 (2½)		x	x	x	x	x	x		
75 (3)		x	x	x	x	x	x	x	x
100 (4)		x			x	x		x	x
150 (6)					x	x		x	x

When sawn softwood is to be used, it is most sensible to use only these sizes whenever that proves possible. Other sizes can be produced by sawing these standard sections, but the price paid is likely to be that of the original section plus the extra sawing costs.

As has been said, hardwoods are converted so as to gain the maximum amount of usable timber. The sizes are, then, much more random, although there is a tendency towards standardization within each geographical source. Thicknesses are as given above for softwoods, with widths varying between 100 mm (4") for the softer hardwoods to 600 mm (24") from some very large, tropical trees.

If two pieces of timber are to be sawn from one section, then approximately 4 mm (⅙″) will be lost to the saw blade. If the timber is to be planed (wrot) then 2 to 3 mm (¹⁄₁₂ – ⅛″) will be lost from each planed face.

(vi) Split timber. Much landscape work can be done with unconverted and unseasoned timber, particularly where a more rugged character is acceptable or desired. The timber should have the bark removed, as this both traps water and encourages fungi and insects. If roughly straight sides are wanted to allow simple fixings, or if the round section is too large for a particular purpose, then the section should be split rather than sawn. The splits will run with the grain, and so not cut through so many cell walls. If a naturally resinous timber, such as fir or larch, is used in this way, a ten year life might reasonably be anticipated.

2.6.2   Preservation

All timbers have a natural resistance to deterioration, but this varies considerably between species. As examples, teak should last for 25 years or more, oak and yew for 15 to 25 years, fir and larch for 10 to 15 years, pine for 5 to 10 years, and birch, beech and sycamore for 0 to 5 years. Some timbers have a much longer life in some environments than they do in others. Elm is a good example. If used where it will be exposed to variations in the weather, its durability would be limited to 5 to 10 years, but if constantly wet (such as when retaining the bank of a canal) it would be expected to last for 20 to 25 years, if not longer.

All timbers can have their natural resistance to deterioration improved by chemical treatment. The more dense the timber, the more difficult this will be to achieve, but the less will be the need to do so. Three main groups of chemicals might be considered.

Traditionally, timber was preserved with tar oils such as creosote. These were effective but the material never really dried and was always likely to smell, stain and burn the skin, and was impossible to paint. Modern preservatives make tar oils obsolete.

(i) Organic solvent preservatives. These have naphthalenes and metal naphthenates, chlorophenols, metal oxides, and the like, dissolved in non-volatile petroleum fractions. To this brew might be added insecticides, water-repellants and other specifics. The precise chemistry is unimportant (indeed, is not frequently disclosed by the manufacturers) but the designer should understand that the timber can be protected from the weather, and from attack by micro-organisms, fungi and insects, such as to give a section of softwood, by way of example, a useful life of 20 years at a minimum. Organic solvent preservatives can be colourless, may have a pigment added as an indicator that the material has been treated, or may have a

decorative pigment added so that no further treatment is necessary (but these are all various shades of brown). Timber that has been treated with an organic solvent preservative can be stained as a finishing operation or might be painted once all of the solvent has evaporated. The instructions/advice of the manufacturer of the preservative should be followed both in respect of the preservative treatment itself and any subsequent decoration. In particular, it should be noted that it is possible for there to be chemical incompatibilities between different treatments.

(ii)   Water-borne preservatives. These have various heavy metal salts (fluor-chrome-arsenate-dinitrophenol is a wonderfully polysyllabic example) dissolved in water. Chemical reactions between these salts, as the water evaporates, give residues which are no longer water-soluble. These provide protection from the agents described above. Water-borne preservatives generally leave a greyish-green colouration to the timber. Some may have a decorative pigment added. All can be stained with chemically compatible products, or could be painted after the timber has been given a stiff brushing to remove any surface salts. Water-borne preservatives are generally more appropriate as a treatment for timber to be used externally than are those with organic solvents, particularly when the work is nearer to carpentry than joinery, and the best specifications would give a piece of softwood in constant contact with the ground (such as a fence post) a life expectancy of 40 years.

(iii)   Application. There are a number of ways in which the timber can be treated. The descriptions below start with the least effective and finish with the best.

(a) Brushing and spraying. Preservatives are available in small cans and might simply be hand brushed on to the seasoned timber. Alternatively a mechanical spray might be used. But it must be stressed that these approaches are not recommended as the sole treatment (little penetration is achieved) but as additional work to pretreated timber that has to be cut or drilled as a part of an on-site operation.
(b) Dipping. Better penetration is achieved if the timber is immersed for several hours in a tank of preservative. This treatment might be considered for kiln seasoned timber with a low moisture content which is to be used, in small quantity, externally.
(c) Pressure impregnation. By far the most effective treatment is for the preservative to be driven deep into the timber by a vacuum/ pressure process. The converted and seasoned (and, possibly, partially worked) timber is placed in a sealed tank. A partial vacuum is induced, and held for a period which varies with the quality of the timber and the required life of the treated timber. The tank is then flooded with preservative and left, under

pressure, for up to 60 minutes. All, or part, of this treatment might then be repeated. By controlling times, positive and negative pressures, and solution compositions and strengths, treatments can be varied as appropriate for different timbers and/ or different end uses. Designers should, then, specify in terms of the required performance rather than attempting to describe in detail the process to be employed; all after reference to manufacturers' technical literature.

2.6.3 Principles of timber fence design

As with all other matters considered in this book, understanding the ways in which timber fences can be constructed is vital to their design. This is conspicuously the case with fences for the ways in which they are put together will, very frequently, be a conspicuous aspect of their appearance.

The functions which a fence might have to satisfy can be defined as (a) to identify a boundary, (b) to prevent the movement of vehicles, people or animals across a boundary, (c) to provide a screen against the weather, or (d) to provide a screen for privacy. The height and density given to a fence will depend upon the combination of these functions applicable in each case.

The major forms which a timber fence might take can be described as (i) post and wire, (ii) post and rail, (iii) horizontally boarded, (iv) vertically/diagonally boarded, or (v) panelled. As should become clear, some of these forms associate more directly with certain functions than do others. In the sections which follow, each of these forms will be considered in turn, starting with the simplest.

(i) *Post and wire fences.* Hundreds of kilometres of simple post and wire fences can be found marking (or incorporated into) field boundaries in all but the very stoniest regions. Whole or split timber posts some 1 500 mm (60″) long are driven (most likely by a tractor-mounted implement) 600 mm (24″) or so into the ground. The posts are de-barked to avoid providing places which might harbour insects or water, and they are pointed to make driving them vertically easier. (100 mm (4″) diameter posts at 3.0 m (10′0″) centres is adequate in most instances. Alternatively, 150–200 mm (6–8″) timber might be quartered for use as intermediate posts. At intervals along the fence every 5 to 8 bays, say, depending upon circumstances, at all changes of direction, and at gateways and the ends of the fence, stouter posts are needed. Indeed, it may prove desirable to brace these stouter posts on one or both sides. This is to take the strain imposed by the wires (Figure 2.31).

The wires may be no more than 3 or 4 strands of galvanized steel wrapped around and twice stapled to end posts and once stapled to intermediate posts as they pass. Where the fence is to be stock-proof, the wires may be replaced with wire netting with a mesh size

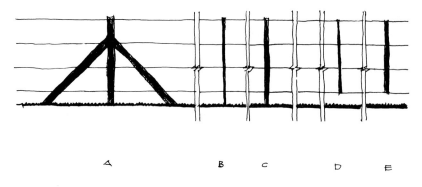

appropriate for the kind of stock. Alternatively, prick posts may be introduced. A prick post is, say, a quartered 100 mm (4″) diameter post set at the centre of each bay and driven no more than 150–200 mm (6–8″) into the ground, to which the wires are stapled. Prick posts ensure that the wires act as one should a cow (or whatever) lean against it. (Light galvanized steel spreaders which clip over the wires, serve the same purpose.)

Clearly, some or all of the strands might be barbed wire. A single strand of barbed wire is often included at the top of a stock netted fence, to reduce the possibility of larger animals doing damage.

(ii) Post and rail fences. Post and rail fences serve much the same purpose as those described above, but are used where something a little more sophisticated, physically and/or visually, is required. Because of this, post and rail fences are much more frequently made of sawn rather than split timber; not only does this look neater but it also allows more positive fixings.

Posts are unlikely to be set by driving as that inevitably causes some damage to their tops. They are more likely to be set in a 300 mm (12″) diameter hole dug with a tractor-mounted auger. If the fence is not to be high, say no more than 1 200 mm (48″), then it should be sufficient to set the posts in as-dug material. That is, the posts are temporarily braced in their desired position and height and the material that was dug out of the hole (other than large stones, obvious rubbish, etc.) is rammed back in layers. Layers should not be more than 100 mm (4″) thick before consolidation, for it is critical that consistency is achieved. Again it will normally prove sufficient if one-third of the overall length of the post is set into the ground – that is, posts for a 1 200 mm (48″) high fence would need to be 1 800 mm (72″) long and set in 600 mm (24″) deep holes. On the same basis the posts to a taller fence would have to be very long. When this is the case, or when wind forces on the fence are likely to be high, setting the posts in concrete should be considered.

In windy conditions, a fence post has to be able to resist essentially horizontal forces. The post tries to pivot at ground level and if it is not to move then overturning must be resisted by the material to the side of that part of the post which is below ground. This is illustrated by Figure 2.32. Setting a post in concrete increases the side area of its below ground length by a factor of about 3 and so increases the area of the substrate providing resistance by the same proportion. This being so, the depth of the hole could be reduced from one-half to one-third of the height of the fence – that is, posts for a 1 800 mm (72″) high fence would need to be 2 400 mm (96″) long overall and set in a 600 mm (24″) deep hole if that hole were back-filled with

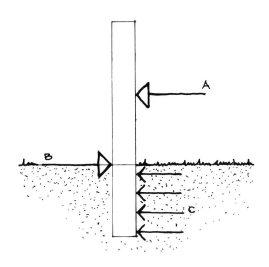

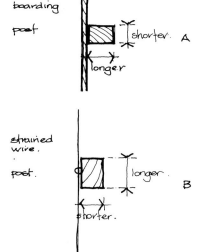

2.32 *A boarded or panel fence will act like a sail. In a wind (A) the post will attempt to pivot at ground level (B). It is imperative that a fence post be set deep enough into the ground that overturning can be resisted by subsoil to the side (C).*

concrete. The concrete does not have to be strong, a 1:3:6 mix would be adequate.

The taller the fence, the windier the site and the higher the proportion of solid to void within the fence, the closer the posts should be set in order to spread wind loads into the ground. As was said when discussing height to thickness ratios of walls, this matter cannot be precisely calculated. A 900 mm (36″) high fence with, say, three 150 × 50 mm (6 × 2″) rails should be entirely sound with 125 × 75 mm (5 × 3″) posts at 2 400 mm (96″) centres. But a 1 800 mm (72″) high fence with six 150 × 50 mm (6 × 2″) rails would need 150 × 100 mm (6 × 4″) posts at 1 800 mm (72″) centres even in only moderately exposed conditions.

Remember that a rectangular timber (or any other) section is best able to resist bending when it has its longer axis in line with the imposed load. So, a fence post which will have to resist wind loads is best set with its longer axis perpendicular to the general run of the fence, whereas with a strained wire fence post (with little wind resistance) the longer axis is best set in line with the fence as a whole (Figure 2.33).

Finally, on fence posts, the tops should always be weathered. That is, they should never be simply cut off horizontally for that would

2.33 *With rectangular sawn posts to a boarded or panel fence (A) it is best if the longer dimension of the post is at right angles to the general run of the fence, whilst with a strained wire fence (B) it is best if that longer dimension is parallel to the general run. Set in these ways, the posts are best able to resist imposed loads.*

193

encourage water to lie and the timber to rot. As a mimimum they should be cut off at an angle of 20–30°. But frequently much can be done for the general appearnace of the fence if more thought is given here. Figure 2.34 shows some examples which might be considered. Not all of the post tops need be finished in the same way – introducing rhythms within the fence or emphasis at gateways, as examples.

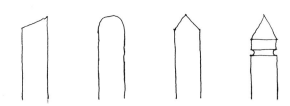

2.34 *The tops of fence posts must always be weathered so that rain-water will run off. This can be done in ways with more, or less, visual impact – an opportunity to significantly affect the character of the fence as a whole.*

Post and rail fences are easy to climb and lean on, and often easy to sit on as well. The rails have to act as beams between the posts. It is this which is most likely to determine appropriate size. A useful rule of thumb suggests that the vertical dimension should be not less than one-fifteenth of the span (rounded up to the next standard size when using softwood). That suggests 150 mm (6″) wide rails with spans up to 2 250 mm (90″), 125 mm (5″) when posts are at up to 1 800 mm (72″) centres and 100 mm (4″) for spans up to 1 500 mm (60″). Clearly wider rails might be used to make a visual point. The shorter dimension of the rail section is best kept between one-third and one-half of its longer dimension.

As with the tops of the posts, the upper surface of the rails should be either splayed or rounded so that water will not lodge. There might, then, be visual argument for splaying or rounding the lower surface also.

Rails might be fixed to posts by planting or jointing (Figure 2.35). A planted rail is simply fixed to one or other face of the post. All the rails might, but might not, be fixed to the same face. The simplest fixing would be a pair of skew nails. Ideally for strength, the nail length should be 2½ times the thickness of the rail, but that can be limited by the thickness of the post and, more often, by the fact that nails over 100 mm (4″) are not readily available. Two nails are the minimum, to resist pivoting action. Two nails are also the maximum, to avoid splitting the timber. Figure 2.36 shows that these should be set as far apart as possible and also shows that they should not be driven perpendicular to the face of the rail but at an angle relative to each other (skew nailing) so that, together, they hold the rail firm. Rails at end posts might be screwed for greater rigidity.

Mild steel nails and screws will rust. Whilst it would take some time before the fixing failed, the rusting would certainly disfigure a painted fence and also a fence stained anything other than rust

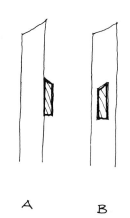

A   B

2.35 *The rails to a fence might be planted (A) or mortised (B). Note that the top of the rail should be weathered in either case.*

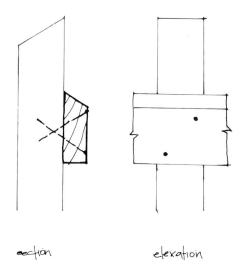

section   elevation

2.36 *A planted rail should be skew nailed to the post. The two nails, at an angle to each other, work in combination.*

colour. Galvanized nails and sheradized screws should be considered.

Clearly rail timbers are not available in unlimited lengths. When posts are set at wider intervals, it might not prove possible for the rails to span more than one bay. By far the easiest way of resolving this difficulty is not to try to give the rails the appearance of being continuous. Alternating the rails back to front, bay to bay, can effectively articulate the fence. Setting the rails at different heights in different bays can do likewise, and this can be particularly useful on sloping ground. When it is important that the rails appear to be continuous and post faces are quite narrow, then ends should be scarfed. Scarfing involves cutting the ends of the rails at an angle, but with blunted points so they do not splinter. This allows the ends of both rails to project across the full face of the post so that each can be twice fixed. Figure 2.37 illustrates these points.

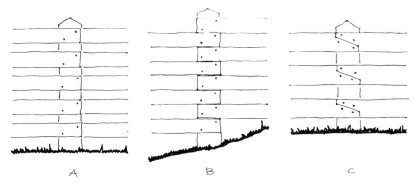

A   B   C

2.37 *Rails cannot be continuous. Rails might be planted on alternate sides of the post (A) or, on sloping ground, they might be fixed at different heights (B). When it is important that they are fixed at the same height and on the same side of the post, they should be scarfed (C) in order to allow an adequate fixing.*

Jointed post and rail fences involve cutting mortices in the posts. A mortice is a slot cut right through the post. In these cases, the mortice would have the same cross section as the rail (plus a small tolerance for ease of fit). Rail ends are frequently scarfed to give adequate bearing within the mortice. Because the rails are held captive within the posts, they do not have to be fixed in any other way. The mortices in end posts should be blind, that is, not cut right through the post, again so as to hold the rails captive (Figure 2.38).

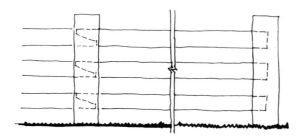

2.38 *Rails cannot be continuous. Mortised rails should be scarfed within the thickness of the post. Mortices in end posts should be blind.*

(iii) Horizontally boarded fences. A horizontally boarded fence is essentially similar to a post and planted rail fence. The only real difference is an increase in the number of rails; this can lead to changes in the sections of those timbers, particularly when the fence is to have no voids.

So, a fence such as is shown in Figure 2.39 should be designed and constructed exactly as described in the first part of (ii) above. But, if boards were made much wider than the span of the rails required they might also be made a little thinner. And if the boards became very wide in proportion to their thickness, they might be three times nailed or screwed to the posts to help avoid them distorting.

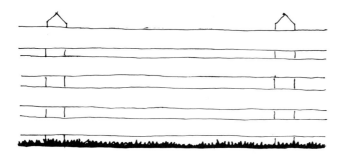

2.39 *An open horizontally boarded fence is essentially the same as a planted post and rail fence.*

A fence which is horizontally close boarded (that is without gaps between the boards) might simply allow rectangular sections to butt one to the next. However, this can produce problems, not the least being that the boards will inevitably move giving an irregular surface and (possibly) loss of privacy. Profiled boards, of which examples are shown in Figure 2.40 would be more satisfactory. Shiplap boards can still move but because the surface is not intended to be flush, that movement does not adversely affect appearance. Weatherboards are housed into each other at their top and bottom edges and so opportunities for boards to move independently are limited. Tongued

196

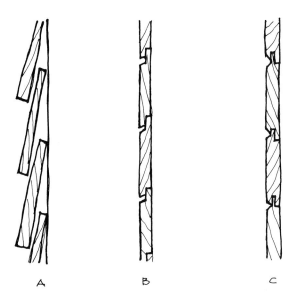

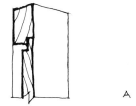

2.40 *A close horizontally boarded timber fence might use boards of different section – shiplap (A), weatherboard (B), tongued, grooved and V-jointed (C) are examples.*

and grooved boards are positively located to their neighbours and independent movement is impossible. But it should be appreciated that these boards will expand and contract in their width and in so doing are very likely to expose lines of unfinished timber at their joints. Repainting and restaining will be necessary after any initial shrinkage. It should also be noted these these boards are quite weak individually and so the fence may be rather vulnerable at its top. Figure 2.41 shows two examples of how this problem might be resolved.

(iv) Vertically and diagonally boarded fences. Horizontally boarded fences do not need rails because the boards can be fixed directly to the posts. With vertically boarded fences this is not possible. Nor is it possible at at least one end of diagonal boards. With these fences rails must be included to give a fixing for the boards. The sizes of the rails should be as already described. Their vertical centres would depend upon the thickness and types of boards to be used. Two rails for fences up to 900 mm (36″) high, three for fences up to 1 350 mm (54″) high and four for fences 1 800 mm (72″) high are common.

Boards are simply nailed once (if narrow) or twice (if broader) to each rail. Tops should not be cut off flush, unless protected by a weathered cap.

Vertically and diagonally boarded fences have far more potential for variety and individuality than do others. The tops of the boards might be cut in a number of ways – and all do not have to be the same. Board widths and lengths can be varied within a fence – modulating the progress of the fence as a whole and/or elaborating the basic appearance at gateways, corners or where fence meets building. Rails might be planted on alternate sides of the posts, and/or boards planted on alternate sides of the rails – with both these devices

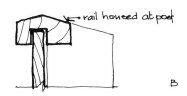

2.41 *The top of a close horizontally boarded fence must be protected – a stouter top board (A) or a capping rail (B) are possibilities.*

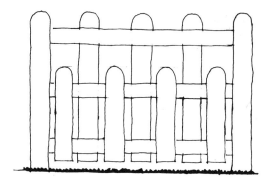

avoiding the fence having front and back sides. The posts themselves can be visually integrated into the whole – be that so as to have a very similar appearance to the boards, or in order to contrast. Figures 2.42 and 2.43 show examples. One further point needs to be made about fixing diagonal boards specifically. If they are to be fixed to planted rails, then a vertical fillet of timber must also be planted on the face of end posts in order to allow a fixing for those boards which do not span between two rails. When diagonal boards are to be fixed to morticed rails, additional fillets of timber (or rebates formed in the posts themselves) are needed at each post, as illustrated by Figure 2.44.

(v) Panelled fences. The above notes are specifically intended to assist with the design of fences with a degree of individuality. But there are also other types of timber fences which use prefabricated panels as their major element. These are made by a number of different manufacturers and so display some detail differences. However, they are all very similar in principle and so can be discussed by generic type. Designers should know and understand the potentials of these fences. It is not always the case that individuality is essential and reference to details used over a wider area than an individual site can often be useful. But, as with all things, lack of thorough understanding leads to misuse.

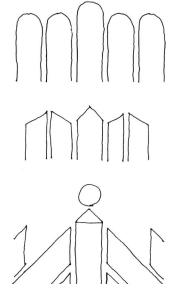

2.43 *The tops of fence posts must be weathered. With vertically and horizontally boarded fences, fence posts can be weathered in ways which augment the characters of the fence as a whole.*

Wattle hurdles. Hazel and willow hurdles are one of the lightest and cheapest of solid barriers, and have the advantage of being easy to put up and take down. This makes them particularly useful for temporary fences, such as providing windbreaks for young planting – which is not to say that they might not also be used more permanently. Their pleasant colour and the texture of the weave goes well with many situations. Whilst carrying the qualities of folk tradition, their straightforward design and construction avoids craftiness. But hurdles should not be used where urbanity is required.

Good quality hurdles should have a life expectancy of 10 to 15 years. They can be made in panels, or woven *in situ*. Panels are usually 1 800 mm (6'0") long and with heights of 900, 1 200, 1 500 and 1 800 mm (3, 4, 5, and 6'0"). Stakes with a diameter of 75 mm (3") for

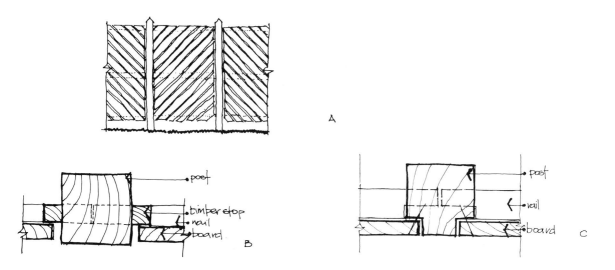

A

B

C

fences up to 1 200 mm (4′0″) and of 100 mm (4″) or thereabouts when taller, are driven until firm and the hurdle fixed to the posts with galvanized wire. The twisted ends of the wire should be tucked into the hurdle to avoid the possibility of scratching. Some authorities recommend that cleft posts should be used (giving flattish sides) and the hurdle fixed, clear of the ground, with a batten that is placed vertically over the hurdles and then nailed through to the post (Figure 2.45). When woven *in situ*, the verticals within the hurdle (whips of about 25 mm (1″) diameter) are driven directly into the ground. No posts are needed unless the fence is to be tall, when stouter stakes are driven at 1 800 to 2 400 mm (6–8′0″) centres.

Hazel whips are split prior to weaving and give a more robust character than do willow whips (osiers) which are more flexible and have a much smaller diameter, and so are used whole. It should be noted that it is not at all unusual for a willow hurdle woven *in situ*, or fixed with the panel in contact with the ground, to take root and turn itself into a hedge. This might be encouraged or discouraged, as suits the designer's purpose.

Interwoven and lapped panels. Lightweight panel fences have become increasingly popular, largely due to their relative cheapness. They are not without potential but are, unfortunately, too frequently used in ways which are, frankly, pretentious. Pretension can be avoided by straightforward detailing which lays stress on the fence as a whole, that is, without allowing the posts to project above the panels, which can look horribly fussy.

Each panel consists of a frame 1 800 mm (6′0″) long by 600, 900, 1 200, 1 500 or 1 800 mm (2, 3, 4, 5 and 6′0″) high. Typically, the frame is made of 38 × 19 mm (1½ × ¾″) softwood, with horizontal slats 100 mm (4″) or thereabouts in width and perhaps as little as 9 mm (¼″) in thickness, which are held in place with 25 × 19 mm (1

2.44 *Diagonally boarded fences cannot be fixed solely to the rails. As shown in elevation (A) the ends of the boards by the posts also need a fixing. This might be done by planting a stop to the side of the post vertically between the rails (B) or by reducing the thickness of the rail where it is mortised into the post so that the post can be rebated (C).*

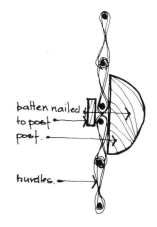

2.45 *A hurdle fence is best fixed to a cleft post by nailing between the hurdles through a timber batten. Nailing through the hurdle could split the whips and/or pull the weave apart.*

× ¾") battens nailed through the slats and into the frame. Larch is a popular material to use, both because of its naturally resinous qualities which give it durability, and its flexibility in thin strips.

Three patterns are common. Interwoven panels include vertical slats which can be thought of as the warp to the horizontal slats' weft. Over-lapped interwoven panels are very similar except that the horizontal slats have shallow recesses cut in alternating sides so that, after weaving, each edge of each slat overlaps either in front or behind, those above and below. Lapped panels are prefabricated feather-edged close boarded fences. Slats with a waney lower edge are often used in lapped panels. Square posts from 63 to 100 mm (2½ – 4") side length, depending upon the height of the fence, are set in any of the ways described earlier in this section. (Some manufacturers will also supply metal spikes with a square cup to take the post, which are simply driven at the appropriate locations. These are not cheap, nor are they technically sound in all locations (particularly for a tall fence) but are very convenient for those without confidence in their manual skills.) Panels can be simply nailed to either the sides or the face of the posts. However, the panel frames have very small sections, and so a more satisfactory detail would be to nail through a 75 × 19 mm (3 × ¾") softwood section for the height of each panel and in line with each post. Figure 2.46 shows that these nails go between the panels, rather than through their frame.

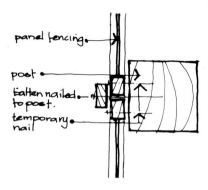

2.46 *Fencing panels are not made with very substantial timbers. So, as with hurdles, fixings are best made by nailing through a batten between the panels.*

Chestnut paling. Sweet chestnut by its nature splits easily into reasonably straight pales with a roughly triangular section of approximately 38 mm (1½") side length. These are given a blunt point at each end and made up in 15 m (50'0") rolls, 600 to 1 800 mm (2–6'0") wide, in 150 mm (6") increments. The chestnut pales are held at 50 to 125 mm (2–5") centres, with two or more double strands of galvanized wire. Each pair of wires are taken one each side of a pale, then given a twist, then one each side of the next pale, and so on. Fences up to 1 200 mm (4'0") have two pairs of wires, and those over 1 200 mm have three.

Chestnut pale fences are fixed to posts as described for post and wire fences. An alternative approach, which gives a stronger fence, is to use cleft timber for the posts, with planted rails. The chestnut paling is then also stapled to the rails rather than solely to the posts.

Chestnut pale fences are most commonly used temporarily. They give good physical protection and act as reasonably effective wind-breaks which slow the wind but without causing eddies. They give little or no visual privacy but when soundly constructed and well maintained, should give 20 to 30 years of life, and so do have potential for more than temporary protection.

2.6.4 Finishing timber fences

Fences which have not had a decorative pigment added during the preservation treatment, might be finished in one of two ways – staining or painting. Both sawn and wrot timber might be stained, but wrot timber must be used if the fence is to be painted.

It is beyond the scope of this book to attempt to describe the chemistry of paints and stains. The matter is very complex (increasingly so with the use of synthetic resins and polymers) and, in any event, is not critical to an understanding of how timber should be prepared for finishing – and if that preparation is not properly done, then the staining or painting is likely to be a failure, however carefully manufacturers' instructions have been followed.

(i) Preparing timber for painting. As has been said, only wrot timber can be given an effective paint finish. Sawn timber has too many surface irregularities which would puncture the paint film, leading to its rapid deterioration. The smoother the timber surface is, prior to painting, the better will be the ultimate paint film. The planed timber should, then, be sand-papered, possibly twice, with a finer paper being used on the second occasion. It would be useful if any sharp arrisses are gently rounded when this is done. It must then be well brushed to remove any loose material and any surface salts which might have been left after preservation (see section 2.6.2). Any major irregularities in the timber surface should be filled with one or other of the proprietary products made specifically for this purpose, having checked that they are not offered solely for internal use. Filled areas should be re-sanded to be smooth and flush. Knots in timber tend to exude chemicals which stain the paint immediately above them. This can be avoided by painting over the knots (during general preparation) with knotting. This is shellac dissolved in methylated spirit (or a synthetic equivalent) which, on drying, seals the knot.

All of the above is written on the assumption that the timber to be painted is new. If it is old timber, but has not previously been painted, then it might be necessary to wash it thoroughly with detergent to remove any oily or greasy dirt, and then rinse it equally thoroughly with clean water to remove any loose dirt, and the detergent.

(ii) Paint. Essentially, a paint is a pigment held in a binding medium. There may also be thinners to give sufficient initial fluidity. After application the paint undergoes changes which leave a thin but flexible film over the surface of the timber. The nature of these changes varies with different types of paint. Some harden due to the evaporation of a solvent thinner. Others because of the evaporation of water from an emulsion. Some (air drying paints) experience oxidation after evaporation of the thinners giving a film which is no longer soluble in those thinners. Finally, some paints harden chemically through the addition of an agent stirred into the paint immediately before use. Three types of paint are needed to finish timber (irrespective of the classes as described above).

201

Primers are high in thinners and are used to penetrate the surface of the timber and to provide a link with the treatment which is to follow. They are often palely pigmented in order to show that all the timber has been treated, and that all has been covered by the subsequent coats. Primers with an aluminium content are particularly useful for landscape work using timbers with a high resin content.

Undercoats are formulated to have high obliteration and body, and so will mask slight visual and physical irregularities in the timber surface. They are pigmented to be almost, but not quite, the same colour as the finishing coat(s) – again so as to give visual signals to the painter.

Finishing paints vary in their chemistry so as to harden with different degrees of surface gloss. The range covers high gloss at one extreme to matt at the other.

Timber is generally given four coats of paint. The primer plus either two undercoats and one finishing coat, or one undercoat and two finishing coats. The latter is more expensive in both labour and materials but is considered to be a superior specification. Each coat should be allowed to thoroughly dry before it is lightly rubbed down with fine sand or glass-paper, so as to provide a mechanical key for the next cost. This is of particular importance with a specification requiring two finishing coats of high gloss paint.

Paints are available in every colour, hue, shade and tone imaginable. Whilst it used to be that some pigments were more likely to fade than others, this is no longer the case. Because of this, designers might feel that they should specify colours which are lower in chroma from the outset than might have been done in the past.

A properly prepared and executed four-coat paint film should last up to six years in a sheltered or moderately exposed environment. In more rigorous situations such as at the seaside, in industrial areas, or in the mountains, a half of that lifetime is all that can reasonably be expected. It is critically important that painted timber be maintained before the paint film shows signs of mechanical breakdown. If that should happen, all of the loose paint must be laboriously removed. Indeed, it might prove necessary to strip (mechanically, with chemicals or with heat) the whole of the original paint film if a regular surface is once again to be achieved. If the original paint film has not physically deteriorated, it should be washed down with detergent, thoroughly rinsed, lightly rubbed down with glass-paper and repainted with one undercoat and one finishing coat, or two undercoats and one finishing coat, or one undercoat and two finishing coats, if the colour is to be changed.

(iii) Stains. Whilst it is a little simplistic, it is not too unreasonable to consider stains as being paints with little or no body. Some simply have the pigment dissolved in a volatile solvent which, after that solvent has evaporated, have done nothing other than, literally, stained the timber. Most are more sophisticated and may also offer a

Rails and open fences can define physical space in a range of ways and for a range of purposes – but always invite the view beyond.

degree of protection from the weather, through the inclusion of silicates, for example. Some manufacturers urge the use of two of their products in combination, the first to adjust colour and the second to afford protection. Some include fungicides and the like, so as to also give protection from biotic agents.

Some care is needed when selecting stains for timber which has been treated with preservative. It is not impossible for the two to be chemically incompatible. Checks should be made with both the manufacturers of the preservative and of the stain. Timber for staining should not be knotted or stopped (filled), for the stain finish will not obliterate those treatments. Timber with a naturally low resinous content will maintain a stained finish for longer before needing maintenance. Similar periods before maintenance can be expected as with paint, although it should be added that faded (but otherwise physically satisfactory) stained finishes, do not demand retreatment as emphatically as do similar paint films. This might be a short-term advantage, but can lead to difficulties if, in consequence, necessary maintenance has been ignored.

Stains are available in a much narrower colour range than are paints. Browns are well covered, from almost black to very light. Bright colours are impossible – the absence of body means that the colour of the timber plays a major part with stained finishes. High chroma colours, greens, blues, reds and yellows, are available but all are essentially dull rather than bright versions of those hues.

Maintenance generally requires no more than a stiff brushing to remove any loose material, and the timber then given one further coat of the original specification. But do take care of any neighbouring plants. They could be injured or even killed, not merely disfigured, by sloppy workmanship.

2.7 METAL FENCES

Comprehensive notes covering the full range of types, classes and varieties of metal fences that are available, would be beyond the scope of this book. All are manufacturered off-site and, in most cases, the required design skill is, primarily, one of appropriate selection. The necessary understanding is of the range of available options, and of the general characteristics of each of those. So, the notes which are offered do no more than give a broad description of each of the major classes. Once a general decision has been taken about the class of fence that is to be used, then further reference to manufacturers' catalogues is critical. These will show the ranges of patterns and finishes that are available, and advise on fixing details and subsequent maintenance. In many cases, direct discussion with possible suppliers should follow prior to a final decision being taken.

The elegance inherent in so much traditional ironwork lies partly in the understanding with which those designers exploited the relative strength of light sections of the material, and also in the discipline they displayed when confronted with a substance which could so easily be formed into almost any imaginable shape. During the late eighteenth and early nineteenth centures, when those qualities were exploited to the full, it was not only the one-off special in wrought iron, but the run-of-the-mill castings mass-produced by the foundries, that enhanced streets and gardens to such effect. The decline in the design of the more architectural of these patterns is only too well known. The twentieth century showed further decline with, apparently, almost all decisions being informed only by the criteria of utility and economy, or by misplaced romanticism. The situation is beginning to improve again with a small number of manufacturers of all kinds of metal fencing offering products which show the same kind of sensitivities to materials, processes, products and applications as was the case one hundred or more years ago. They should be encouraged, and others encouraged to join them, by landscape designers showing themselves capable of selecting and using metal fences that are entirely right in their context, however grand or humble that might be.

Metal fences have most potential where a physically and psychologically strong barrier is needed, but one which gives little visual obstruction, be that in the town or country. Vertical iron railings, decorated to a greater or lesser degree, have become typical of our cities, suggesting an urban character, and netting the view by giving the public glimpses or vistas of places to which access is either limited or prohibited, whilst at the same time protecting that property by their very toughness and spikiness. By contrast, plain rail (or continuous bar) fencing has become an accepted part of the rural landscape where the larger scale demands a more straightforward response. The clarity of the line effectively defines, say, lawn and paddock, whilst its lightness both opens up the broader views and acts as a foil to heavy summer foliage or winter's complex tracery. Similarly, the more sophisticated forms of wire mesh fencing work better in the town than they do in the country, whilst in either location it is critical that their near invisibility is recognized and exploited.

2.7.1 The materials

(i) *Ferrous metals.* Few 'iron' fences are still made of iron, but of an alloy of iron with carbon (steel) and other elements.

Ferrous metals with a high carbon content melt at a lower temperature than do others. This makes them particularly suitable for casting. The higher carbon content also gives some protection against rusting. Those with a lower carbon content (often known collectively as mild steels) can be forged (worked whilst hot), cold-worked and (when the proportion of carbon they contain is very low) drawn into wires. But steels with little carbon are very prone to rusting. Stainless steels contain a proportion of chrome and on exposure to air these develop an invisible film which is highly resistant to oxidation. Weathering steels contain copper. These develop a purplish-brown coat of rust, but one which then protects the material beneath. However, during the 18 to 24 months which it takes for this coat to develop, they are very likely to stain any surfaces below them.

(ii) Aluminium alloys. Pure aluminium is not suitable for metal fences, but when combined with magnesium, manganese, silicon, and others, it can be cast, forged, drawn or extruded, depending upon the particular alloy. Aluminium alloys become a dull grey when exposed to the atmosphere, but do not otherwise deteriorate unless subjected to electrolytic action or attacked by acids.

2.7.2 Joining metal components

(i) Welding and brazing. Aluminium alloys and low carbon steels can be joined by welding. This involves locally melting adjacent surfaces so that they fuse. Often additional material of the same metal as the parts being welded is introduced as a strengthening fillet.

High carbon steels cannot be welded for they cannot be subjected to the necessary intensity of heat. They can be brazed (as can aluminium alloys). Brazing involves the introduction of a second metal which has a lower melting point than that of the casting. Brazing has much less strength than has welding.

(ii) Mechanical joints. Metal components with some substance (that is, excluding wire) can be mechanically joined with bolts. Mild steel bolts are available in a range of lengths from 20 to 300 mm (¾ – 12″) and with a range of shank diameters of 6 to 20 mm (¼ – ¾″), but only the smaller of these would have a use with metal fencing.

Most bolts are combined with washers below both the bolt head and the nut. The washers are important in order to spread the load induced by tightening the nut. A neater fixing is given when the holes have been tapped (given an internal thread) to take the bolt directly. A washer is still needed below the head, but a second washer and separate nut is not required.

Aluminium bolts are not common, even for use with aluminium fences, for the material is so soft that threads strip when the nut is tightened. But bare mild steel bolts should not be used with aluminium due to the risk of electrolytic action. Bare mild steel is not, in any event, a good idea because of the problem of rusting.

Metal components which are flat and relatively thin can be joined by cold rivets. These consist of a plug of ductile metal fitted over a pin which is inserted into the nose of a rivet gun and the plug then inserted into the aligned holes. The gun pulls on the pin and the soft plug spreads on both sides of the components to be joined. The gun cuts the pin off flush when the rivet has been successfully placed.

Rivets have a very neat and tidy appearance, but again the manufacturers' advice must be taken to avoid the risk of electrolytic action.

Wire fences are often fixed by the simple expedient of twisting. Fixing a straining wire with staples to a timber post has been described in section 2.6.3. Clearly, a wire cannot be stapled to a metal post. At the minimum there must be a pre-drilled hole through which the wire can be passed prior to twisting it around itself. This is not a tidy detail, nor is it all that secure. A more satisfactory detail threads one or more soft metal sleeves over the wire before it is passed through the prepared hole. The wire is then passed back through the sleeves which are then squeezed tight with a special tool. To strain the wire it is fixed, as above, to an eye-bolt set in the post. Tightening the nut to the eye-bolt tightens the wire. The projecting length of thread might then be taken off with a hacksaw – giving a safer detail with better appearance – but that does mean a new eye-bolt would be needed should the wire break at some later date. Some manufacturers supply hollow posts with winders contained within them. This is the most satisfactory detail, both practically and visually. Figure 2.47 illustrates these alternative fixings.

2.7.3   Finishes

(i)   Ferrous metals. Stainless steel and weathering steels need no additional finishing. Whilst ferrous metals with a high carbon content are less prone to rusting, they, and all others, need protection. Shot-blasting or pickling with acid effectively removes existing rust and scale. Wire brushing does this less effectively. Treatment with acid phosphates gives a matt grey-black surface which inhibits the establishment and spread of rust.

Coating with a non-ferrous metal affords protection. Zinc is most commonly employed for this. It might be applied through an electro-plating process (the original galvanizing, although this term is now often used for any zinc coating), by hot-dipping in molten zinc, or by tumbling in heated zinc powder (called sheradizing, but limited to small objects only). All of these treatments are best done after any prefabrication; for any further cutting, drilling or welding must leave untreated steel exposed to the weather. Galvanized steel can be used without any further treatment; but the coating will break down in time, and its appearance is never anything other than dull. Zinc coatings are best, then, considered as preparations for painting – not substitutes.

Paints and painting have been broadly described in section 2.6.4. Those principles apply to painting metals also. Preparation is different (see above) and so are the primers.

The traditional priming paint for ferrous metals is red lead. It is very effective (if a little slow drying) but **must never** be used in situations accessible to children and animals, for it is highly toxic. If

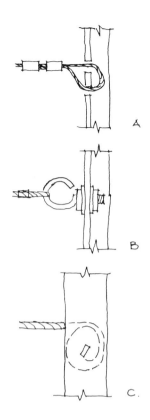

2.47 *Wires can be fixed to metal fence posts simply by twisting, but this both looks shoddy and can be dangerous. The examples shown here are more satisfactory. (A) is a permanent fixing, whilst (B) and (C) allow the wire to be re-tensioned. In the latter case, the mechanism is contained within the post.*

207

the fence has been given a zinc coating, then it will need to be etched so as to provide a key for the paint film. Products are available for this purpose, but are not satisfactory as a sole primer. They are susceptible to moisture and so immediate over-coating is advised. Zinc phosphate primers are most effective on steels, as are those with zinc particles held in a synthetic resin medium (often called cold galvanizing but better described as zinc rich primers). Zinc chromate primers have been particularly developed for aluminium, but are also useful for ferrous metals when a high quality of surface preparation has been possible. So-called universal primers should be avoided – whilst convenient, they cannot be entirely satisfactory for all applications.

Once primed, the fence should be completed with either two undercoats and one finishing coat, or one undercoat and two finishing coats, all as the notes in section 2.6.4. As always, manufacturers' literature should be consulted prior to specification and, if in any doubt, discussions held with the manufacturers' technical experts.

If properly prepared and applied, a paint film on a steel fence should have a 4 to 6 year life – given it is not in a harsh environment such as at the seaside, or in an industrial area. Regular inspection is essential and prompt treatment given should there be any signs of rusting. As with timber, it is much more satisfactory to repaint a metal fence before there are obvious signs of deterioration.

Ferrous metal fences may be supplied with a factory applied finish. These require no further work on site, other than to take particular care to avoid damaging them. Most give better protection than would a paint film, but are impossible to replace once that protection is lost. These include stoved paint finishes and electroplating, such as chrome. Only one other will be specifically mentioned here, for it gives protection where painting would, at best, be extremely laborious.

Chain-link and other wire fences could, conceivably, be painted but that would be excessively time consuming to do with a brush, and excessively messy with a spray. For many years a thin zinc coating was all that could be given to protect such fences. But more recently they have been available fabricated from plastic-coated wire. Only a limited range of colours are available as standard (black and green being most common) but others can be supplied to special order for larger contracts. Some of the early experiments with this finish were not an unqualified success – the pvc used became brittle and split. But the chemistry is now much better understood. If the wire has also been given a preliminary zinc coating (as it always should be) then a life of up to 30 years might reasonably be expected. When plastic coated wire fencing is being used with metal posts and other components, they can also be provided with a plastic coating, giving a consistent finish to the whole of the fence.

(ii) Aluminium. Aluminium which has been given no particular finishing treatment will weather to a dull whitish-grey but, unless

exposed to a very polluted industrial atmosphere, will not seriously deteriorate. For some cast components this self-finish might be entirely satisfactory. If not, then the material can be given alternative treatments. If there is any risk of the fence being dirtied with lime-rich droppings (from mortar, or concrete, or birds) then further treatment is recommended, for the lime will leave permanent stains.

Aluminium alloys can be painted, as for ferrous metals. Zinc chromate primers are particularly recommended. They can also be given a range of factory applied paint and plastic finishes – see manufacturers' catalogues.

There is one finish which can be given to aluminium which is not an option with other materials. Anodizing is an electrochemical treatment which gives an integral skin to the aluminium which is harder than the metal beneath. Moreover, the untreated metal has a porosity which allows it to be dyed – gold, black, brown and dark blues and greens have proved to be particularly fade resistant and so suitable for external use. Further, where the nature of the component allows, the surface of the aluminium might be chemically etched or mechanically polished prior to dyeing and anodizing, to give textures varying between matt and mirror. The problem of staining remains and it is advised that anodized aluminium is regulalry washed down with mild detergent – as often as monthly in industrial areas whilst annually should be sufficient in rural locations.

2.7.4 Types of metal fences

The notes which follow will briefly describe a number of different types of metal fencing commonly manufactured. Several of these are covered by a British Standard specification and so there are basic similarities between different manufacturers' products.

(i) Continuous bar fencing (Figure 2.48(A)). This provides a robust but light barrier but not one which would keep out intruders. A series of horizontal flat or round bars (the actual number varies with overall height) are supported at maximum 900 mm (3'0") centres by flat, T or I section standards (posts) with more substantial standards (joiners) at 4.5 m (15'0") centres. Most patterns have the lower bars closer together than the upper – a detail which displays their agricultural parkland origins. Flat bars are a little cheaper than round, but a substantial round bar is recommended at the top of the fence, particularly if it is likely that it will be climbed or used as spectator seating. Standards are generally cut off half-round at their tops, although some have a curve or a crank, which locates the top bar inside the principal face of the fence. Cast terminal pillars are available from some manufacturers, giving a visual emphasis to the ends of the fence, or at gates, etc. Heights of 1 050, 1 200, and 1 350 mm (3'6", 4'0" and 4'6") are commonly available.

Standards are usually driven 450 to 600 mm (18–24") into the ground. On very soft ground, standards with two-pronged feet, or with thrust plates, may be used. They can be set in concrete as with timber fence posts (section 2.6.3) in which case they should be obtained with base plates. Bracing stays are necessary at any radical change of direction. On undulating ground this kind of fence can be built simply to follow those undulations. On a consistent slope the bars can be fixed parallel to that slope. Slight curves on plan can be accommodated with the standard product, or bars bent to specified radii can be supplied.

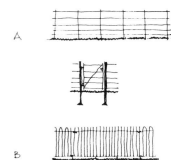

2.48 *Iron fencing is available in a range of patterns and sizes, and with posts (standards) which can either be driven directly into the ground or set in concrete. (A) shows a continuous horizontal bar fence with matching gate. Gate-posts can be more substantial, and more ornamental, and should always be set in concrete. (B) shows a bay of unclimbable fence whilst the neighbouring bays show an alternative which would offer less security but greater safety.*

(ii) *Vertical bar 'unclimbable' fencing (see Figure 2.48(B)).* Square or circular bars are supported vertically at approximately 125 mm (5″) centres with two horizontal rails, one at or near the bottom and the other 150 mm (6″) or so below their top. The height of the lower rail above the ground can be slightly adjusted by varying the depth at which posts are set. Some patterns allow the bars to project below the lower rail, making a more effective physical barrier. They are made up in panels in the order of 2.7 m (9′0″) long. The tops of the bars may be blunt, may be spiked, or may be variously wrought to resemble spears. Alternatively the bars may be paired, to give single or interlaced bow tops. These must always be used where there is any risk of accident. Heights available vary between 900 and 2 100 mm (3–7′), in 150 mm (6″) increments.

Flat or I section standards with a base plate are set in concrete at the centres dictated by the length of panel to be used. Side stays to alternate standards are recommended for taller fences. The panels are then simply bolted to the posts. Ornamental cast pillars are available. Welded vertical bar fencing must be stepped at the standards if the ground is sloping. But self-adjusting patterns are available. These will accommodate changes in slope angle from horizontal to 1:3.

(iii) *Corrugated pale fencing (Figure 2.49).* The vertical pales of this fencing are fabricated from pressed mild steel strip to give a corrugated section 50 to 75 mm (2–3″) wide. They are made up into panels approximately 2.7 m (9′0″) long. Heights vary between 1.2 and 3.3 m (4–11′). The pales are bolted or riveted to steel strip or angle rails which, in turn, are bolted to I section steel posts set in concrete at intervals given by the length of the pre-fabricated panel.

The tops of the pales can be variously cut to semicircular or parabolic curves, with notches, or cut and bent to give triple spikes. But these last patterns should not be used for fences below 2.1 m (7′0″) high or where there is any risk of accidents.

(iv) *Circular and rectangular hollow sections.* Mild steel is available in tubular sections between 25 and 115 mm (1–4½″) outside diameter, and in square and rectangular hollow sections with side lengths between 13 and 125 mm (½–5″). Lengths of 7.2 m (24′) are commonly available, but up to 9.75 m (32′) can be obtained. These can be used

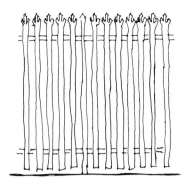

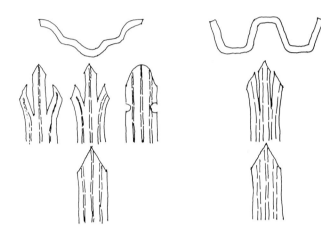

2.49 *Corrugated pale fencing is commercially available in a range of heights and patterns.*

as posts for a wide range of different kinds of fences. They can also be shop welded to form frames, with or without integral legs, which might be used as a simple bar fence or as support for a mesh or pale barrier. There are manufacturers who offer fences made in this way as standard products. Alternatively, designers may choose to design their own. There are also manufacturers who supply fittings for use with hollow sections. These allow tubes to be connected in any number of different configurations, and include curved fittings and stop ends. The range is far too wide to attempt to illustrate them all – manufacturers' literature should be consulted.

(v) Welded strip fencing (Figure 2.50). Metal strip has the particular advantage of being easily bent perpendicularly to its longer side whilst effectively resisting attempts to bend it in the opposite direction. Fencing can, then, be made with this material that has considerable visual interest when seen elevationally, whilst being very strong should forces be applied to the fence as a whole. There are manufacturers making standard fences in this way (most often for use alongside busy roads) but there is a great deal more potential than appears, as yet, to have been exploited. Designs should be evolved

in direct consultation with the workshop. Fences with integral legs, and panels for bolting or site welding to standards, could be considered. This approach would have a particular value when the fencing is to be visually integrated with a particular combination of slopes, or where fencing has to double as handrailing to steps or ramps.

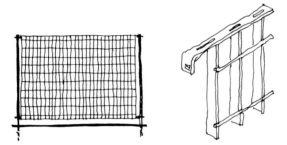

2.50 *Mild steel strip can be welded into panels for fixing with bolts or more welding to steel posts set in concrete. A number of manufacturers offer fences of this kind as standard items.*

(vi) *Welded wire fencing (Figure 2.51)*. Welded metal mesh fabrics are made for a number of different building and industrial applications, not only fencing. A large number of mesh sizes and shapes are available. Different wire sizes are used both between and within fabrics. Different metals allow different finishes or, as with stainless steel, may need no separate finish.

These fabrics can be used in two ways for fencing. They can be incorporated into frames made of strip, angle or hollow sections. When used in this way they would be delivered to site prefabricated as panels for bolting to standards or, if they have integral legs, directly concreting into place. Alternatively, they can be delivered to site in rolls or sheets without a frame, and supported on the posts with straining wires. In addition, it is possible to have the fabric sheets crimped in a workshop, so that the whole has a greater rigidity than is otherwise the case.

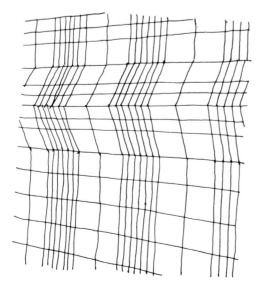

2.51 *Proprietary fences are available made with a variety of welded wire meshes. These, in themselves, have little strength but rigidity can be greatly increased in the way illustrated here.*

212

Manufacturers' catalogues should be consulted for details of proprietary fences using welded mesh fabrics, or the producers of the fabrics should be approached for advice when original designs are being prepared.

Fences using welded mesh can have a degree of sophistication that is not found in woven wire or chain-link fences. They are, then, better suited for urban rather than rural applications. They always look best when fixed to slender standards.

(vii) Chain-link, expanded metal and woven wire fences (Figures 2.52 and 2.53). Light metal I sectioned standards can be used to support a simple strained wire fence. The number of wires used and their vertical centres should vary depending upon application. The standards should be driven (or given a base plate and set in concrete) at maximum 2.7 m (9') centres. Mid-span droppers (see section 2.6.3) might be incorporated.

A strained wire fence of this kind could be used to support woven wire of chain-link fencing. The number of strained wires used will vary with overall height; two for fences up to 900 mm (3') and then at maximum 600 mm (2') vertical centres, is a good rule of thumb.

Woven wire fencing is made with a range of different square and rectangular mesh sizes varying between 75 × 300 mm (3 ×12″) and 225 × 300 mm (9 × 12″). It has been developed, primarily, for agricultural use and so the different patterns are frequently described in terms of the kinds of stock with which they are intended to be used. It is purchased in rolls with widths between 750 and 1 125 mm (30–45″) and lengths of 15 m (50') or longer to special order. Posts are purchased separately, with cleft timber being most frequently used. Posts centres of 3.6 m (12') would be an acceptable maximum – although less would be necessary with heavier stock. No finish other than galvanized is available.

Chain-link fencing is made from zig-zagged wire such that the zigs in one length of wire catch around the zags in the next. The result is a square mesh of 38 to 50 mm (1½–2″) side length, presented diagonally. Roll widths vary between 900 and 2 700 mm (3–7') with widths of up to 3.6 m (12') being made for special applications such as bordering tennis courts. Two edge treatments are available. With the safest, the ends of adjacent wires are hooked around each other to turn back away from the edge. With the other, adjacent ends are simply twisted together and left projecting as barbs. The use of this latter is not recommended for fences less than 2.1 m (7') high. Chain-

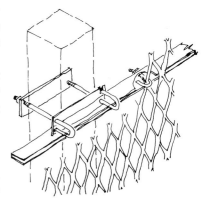

2.52 *Expanded metal is made by cutting myriad little slits opened up into lozenges. This figure shows a commercially available system whereby expanded metal can be clipped to light metal rails which in turn are fixed to timber, concrete or metal posts.*

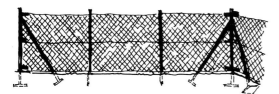

2.53 *Chain-link fencing is held by strained wires fixed to standards. Most chain-link manufacturers also supply the standards and fixings appropriate for various applications.*

link might be fixed to the straining wires with twists of light gauge wire or, much more satisfactorily, with proprietary clips. The posts might be timber, steel or concrete, set as appropriate. Centres might be as much as 2.7 m (9′) for a low fence, reducing to 2 m (6′8″) for a very tall fence. Both galvanized and plastic coated chain-link can be separately purchased but, in the latter case in particular, it is better to order all necessary materials and components from the same specialist supplier or subcontractor.

2.8 OTHER FENCING MATERIALS

This section can be very short for there are no outstanding matters of principle, and manufacturer's technical literature should, again, be consulted for further details.

2.8.1 Concrete

Precast reinforced concrete fence posts have the advantage that they do not rot or rust in the ground. That being so, they can be set with *in situ* concrete at depths and centres appropriate for the type of fence they are to support and, provided that they suffer no mechanical damage, will have an almost indefinite life. They can be cast with an integral cranked top for use where high security is required – a feature which can be given to metal posts but which is only clumsily achieved with timber.

Fences made entirely of precast concrete are available. These may simply be post and rail, may consist of butted or overlapping planks slotted into H sectioned posts, or may be cast as panels that each fill the whole of one bay of the fence. The more utilitarian are simply self-finished whilst others have exposed aggregrates.

For a reason which cannot be adequately explained, precast concrete fences and fencing components always appear rather heavy and cumbersome. There has to be more potential in this technology than has, as yet, been realized.

2.8.2 Plastic

Proprietary fencing is available which uses hollow sectioned plastic for rails, pales and/or posts. These are, generally, offered as low-maintenance substitutes for timber. Perhaps it is for this reason that they rarely seem to make a significant contribution to the design of which they are a part. Cheap alternatives generally cheapen rather than enhance. Good design (of both the parts and the whole) comes when materials are used in ways which exploit all aspects of their nature, rather than solely those ways in which they are better than others. As with precast concrete, there have to be many ways to successfully design plastic fences which have yet to be thoroughly investigated.

2.9 GATES

A gate must appear as the way through a fence or wall. It is part of its function to state visually, however slightly, that here is a possible entrance or exit. If the gate looks stronger than the fence which flanks it, it will, in aesthetic terms, contradict its function. Many gates, sensible enough in themselves, look ridiculous because they ignore this principle. This does not mean that the gate could not be of the same material and construction as a fence of which it forms a part. Some of the most satisfactory looking gates are really a moveable extension of the fencing. It is simply the presence of latch and hinges, and a slight break in the rhythm of the fencing, together with the gate posts, that suggests that there is an opening. A close-boarded gate looks right in a close-boarded fence or in a brick wall, but wrong in a palisade fence; a gate clad in sheet metal looks incongruous in an open-barred metal fence.

For much the same reason, a gate should not be higher than the wall or fence which flanks it. Examples which illustrate this point can be seen where decisions have been taken to use a dwarf wall or low rail to state the existence of an ownership boundary, but where physical deterrence is not needed. A low gate would be a physical nonsense, for it would be so much easier to step over it rather than to stoop to the latch. But a gate at normal height is also nonsense, for why should it deter when no other part of the boundary treatment does so? Better, in such circumstances, to return the wall or rail to the building adjacent to the entrance door, and accept that the pathway is visually a part of the public domain. Indeed, this detail can work well as a welcoming gesture.

But ideas of this sort, when pressed to the limit, often contradict the rule which governs them. Extreme examples of gates which are higher than the barriers in which they form openings, are farm gates at canal crossings in Holland. (Similar examples can be seen in parts of East Anglia, in south-west Lancashire, and on the Somerset levels.) Since the canals are below eye level, the result is a landscape which appears incongruous to a visitor, and which seems to consist of endless rich grassland with cows but dotted with gates placed, apparently, entirely at random. But these examples intrigue rather than jar; probably because they are strictly functional and are consistent over a large scale, and perhaps also because they are so unfamiliar to a visitor.

Further, this note refers specifically to gates and not to gateways. If the gates are set within a masonry or wrought iron arch, for example, then emphasis is on the whole event and not the relative heights of the parts. Indeed, the gateway can continue to make its statement when gates and wall, or fence, or hedge, have been removed – or, even, were never there. Examples as much distanced in time and space as medieval Europe and 20th Century Mid-West America attest to this.

2.9.1 Functional principles

The essential function of a gate is to keep closed, for the majority of time, a gap in a barrier needed to allow people, animals and/or vehicles and machinery occasionally to pass. A gate must be sufficiently robust to resist the forces applied to it by gravity and use. The posts which support it must also be sufficiently robust. It must be convenient to open and close, and it must be possible to keep it closed when not in use. It must be sufficiently wide for the traffic that will use it. The simplest possible form for a gate is illustrated in Figure 2.54. Examination of this example will reveal virtually all the principles which govern aspects of gates.

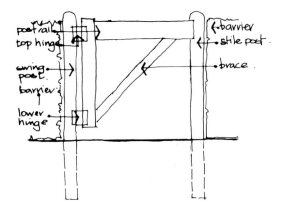

2.54 *A gate in a hedge could not have fewer components than shown here. If the gate was in a fence or wall, the posts might be less obvious.*

Theoretically, the only element of this gate that is needed to satisfy its essential function is the rail; for it is that which maintains the barrier when the gate is closed. All other components are there so as to keep the rail in an appropriate spatial relationship with the barrier, both when the gate is closed and when in use. The top hinge is there to allow the rail to open and close, and the hinge-post is needed to give something to fix the hinge to. The brace is needed to prop the end of the rail and the swinging-post is then required to allow a fixing for the heel of the brace. The lower hinge must then be added so that the swinging-post (and thus the gate as a whole) remains vertical. The stile-post is needed to give the gate something to close against and to be latched to. This gate would be unsatisfactory in many circumstances because animals and small people could simply pass beneath it. Figure 2.55 again shows the simple example, but with a stile, bottom rail and pales added – so resolving that outstanding difficulty.

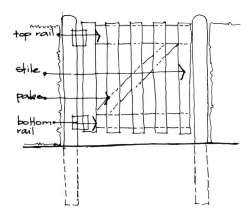

top rail
stile
pales
bottom rail

All gates must have all of these elements or they will be defective. Some, it is true, may appear in other guises – the gate-posts may also be fence posts, or the posts may not be posts at all if the barrier is a wall, for example. It is also true that the dimensions of the components and the detail ways in which they are joined together, will vary considerably with the height and width of the gate and the materials of which it is made; but whatever the details and materials, any well designed gate will, on analysis, show itself to have components which satisfy all these functions.

2.9.2 Widths of gates

The appropriate width for a gate, or combination of gates, will partially be a matter of psychological appropriateness and partially a matter of functional necessity. Little can be said here about the first of these – so much must be judged in the light of specifics. Clearly a gate that is intended to exhibit a degree of pomp and ceremony will tend to be large whilst one intended to be discreet and intimate will tend to be small. Equally clearly, a gate which, due to functional necessity, has to be large will also have a degree of presence, even if that function is relatively humble. We say equally clearly despite the built evidence that this is not entirely clear to all designers.

In regard to functional necessity the following list gives some minimal dimensions for the passage of people and vehicles:

600 mm (24″) for a single person
800 mm (32″) for a pram or buggy
800 mm (32″) for a bicycle
900 mm (36″) for two people (just)
950 mm (38″) for a wheelchair
1 150 mm (46″) for a pram or buggy plus a walking child
1 200 mm (48″) for two people (comfortably)
1 700 mm (68″) for two wheelchairs or prams
2 100 mm (84″) for a small/medium car

2 400 mm (96″) for a large car, ambulance or medium van, or small/
            medium tractor
3 000 mm (120″) for a car and bicycle, or large tractor
3 600 mm (144″) a fire engine, dust-cart, or lorry
4 100 mm (164″) two cars can (just) pass
4 800 mm (192″) a combine harvester, two cars (comfortably), or a car
            and a lorry (just)
5 500 mm (220″) for any (normal) combination of two vehicles.

All of the above dimensions will allow people and vehicles to pass through a gate and also pass each other. But note that clearance is not necessarily simply between gate posts – the open gate may, itself, occupy some of the available space. Note also that dimensions assume straight travel. If a vehicle is also negotiating a bend, then widths can be considerably wider.

2.9.3 Timber gates

(Note: This section should be read together with that on timber, section 2.6).

Entirely satisfactory timber gates can be designed using only planted joints. Figure 2.56 shows such a gate designed and made by the authors. (It was because we were to make it ourselves that more complex joinery was avoided.) Rather substantial softwood was used for the pales so that they could also double as swinging-post and stile. The pales were each twice screwed to both the top and bottom rails. As the pales were so wide and so firmly fixed to the rails, the gate functioned perfectly satisfactorily, without a brace, for a number of years. However, our growing family soon put the gate to use as a piece of play equipment, and immediately demonstrated its lack of inherent strength. But the brace was then added, twice screwed through each of the pales as it passed, and is now entirely adequte for both our and the children's purposes.
    We are quite pleased with the appearance of this gate when seen from the public footpath, but it does somewhat overtly display its

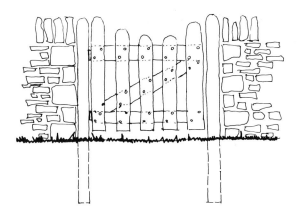

2.56 *This gate was made entirely with screwed planted joints.*

218

more utilitarian aspects when viewed from the garden. This is an almost inevitable problem with timber gates made solely with planted joints. Additional pales could be fixed on the garden side – but these would have to be housed over the hinges, and would add significantly to overall weight. More complex joinery is essential when a timber gate has to be visually attractive when seen from either side. Figure 2.57 shows a simple braced frame which, through the use of mortised joints, is flush on both faces. (Note that the joints are located so as to minimize the possibility of rain-water running into them, and that the upper surfaces of the rails are weathered so as to maximize the possibility of it running off.)

Because this basic frame is flush, it must have the potential to be given pales in a number of different ways, on either or both sides. However, it would not really be satisfactory to complete the design for the frame of a gate, without taking into account the way in which it is to be finished. The design of the frame must be such as will effectively accommodate the pales, and it could be that decisions about the frame would be modified by thoughts about the pales and the gate as a whole. Indeed, it is possible for pales to be introduced in ways which make some of the frame members (the brace in particular) redundant. Figure 2.58 illustrates a variation on the basic mortised frame, where the pales are an integral part of the design. There have to be many more possibilities.

Gates which are wide enough to allow the passage of vehicles, but are also low enough to retain a human scale, can present a particular problem with bracing – for if the brace were to run from corner to corner of the frame, it would be too close to the horizontal to act as an effective prop. There are three ways, in principle, in which this difficulty might be resolved.

First, to borrow a term, the gate might be given a stressed skin. That is, if the whole of the frame were close-boarded in a way that ensured that those boards acted as a single element, then the rectilinear frame could not distort. Those boards would not have to run on the diagonal of the frame, they could be vertical or horizontal, or at any other angle. The strength of modern adhesives gives this

2.57 *A gate frame which is made up with mortised joints would be stronger than the example illustrated in Fig. 2.56, and could be paled in a way which gave a gate which less obviously had an inside and outside.*

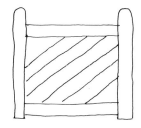

2.58 *Two diagonal pales (also mortised into the frame) have been added to the simple frame shown in Fig. 2.57. There are many other ways in which pales might have been added to that frame, but the design of the whole should be considered before frame details are finalized.*

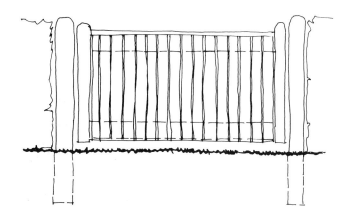

2.59 *The brace might be omitted from a gate if it is finished in such a way that it cannot distort under its own weight. In this example vertical tongued and grooved boards are glued to each other and glued and screwed to the top and bottom rails.*

approach more potential than was the case in the past. Figure 2.59 illustrates one such approach.

Secondly, the brace might be designed to run from the heel of the swinging-post to prop the top rail nearer to the mid-point of its span, rather than where it meets the stile. Figure 2.60 illustrates the principle and shows two traditional gates which variously exploit that principle. Note that with both these designs, the top rail is tapered so that it has thickness where it needs strength, but carries no more weight than is necessary elsewhere.

Thirdly, and this is a more radical departure, the swinging-post might be extended to project above the general height of the gate, and the brace replaced with a tie. Figure 2.61 illustrates this principle with a very elegant traditional form. The tie would be in tension (as opposed to a brace which is in compression) and so could be a quite slender section, indeed, could be a thin metal rod.

Finally, hurdle and light panel timber fencing can be given gates to match provided that they can be incorporated with a braced frame of sufficient substance. A hurdle might be woven about such a frame (a variation on the stressed skin approach!); whilst a light panel might either be made with a more substantial frame, or have additional framework built around it.

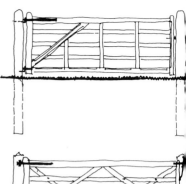

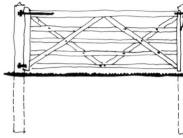

2.60 *A brace running from corner to corner of a wide, low gate would not be very effective. This figure shows two ways in which this problem has traditionally been overcome.*

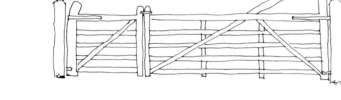
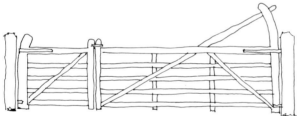
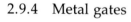

2.61 *Where a brace would be ineffective it might be replaced with a tie.*

2.9.4 Metal gates

(The notes about metal fencing, section 3.7, should be read in conjunction with those which follow.)

The principles, and variations on those principles set out above, apply just as much to metal gates as they do to timber. The differences noted here are, then, ones of constructional and not structural detail. Casting a metal gate as a single piece is not an option that can be seriously considered. The visual qualities which might be achieved in that way can be more conveniently and more satisfactorily matched by other techniques.

Cast components could be included in a gate. So it would be possible to repeat cast details which feature in a neighbouring fence by mechanically fixing them to a wrought and/or welded frame. Figure 2.62 shows how cast decoration might be included within a metal fence and associated gates in a way which makes fence and

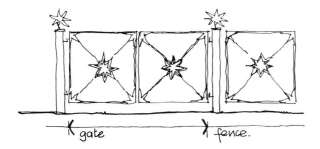

2.62 *A metal gate might be designed with cast embellishments which both disguise the functional necessity of some of its parts and integrate the gate within the fence as a whole.*

gates a single design and disguises the necessity of including braces within the gates.

However, fences and gates could only be made in that kind of way when budgets were high. Generally speaking, metal gates will have a welded frame to which bars, strip, sheet and/or mesh is fixed mechanically or by further welding.

Welded metal gates have one conspicuous advantage over timber gates, and over other metal gates which are constructed with mechanical joints. Properly executed, a weld will sucessfully resist distortion. So, provided that the rails and stiles themselves have sufficient substance, then a welded frame may not need a brace. Metal angles, T sections and channels, and circular and rectangular hollow sections all have the potential to resist much higher bending stresses than could the same weight of metal in a solid bar or strip. This is particularly the case when the longer dimension of a section (when it has one) it positioned vertically within the gate. Gravitational forces act downwards and (as a very simple experiment with a thin strip such as a ruler or scale will confirm) it is much easier to bend such a section downwards when its long dimension is horizontal, than when it is vertical.

An associated disadvantage of the high strength to weight relationships of these sections, is that they are difficult to bend when a gate with curved rather than rectilinear qualities is wanted. Small and thin-walled hollow sections can be bent to quite small radii with a special machine, but even then it is possible to have a mildly crimped effect on the inside of the curves. With tubular sections, this problem might be resolved by using elbow fittings. But with rectilinear

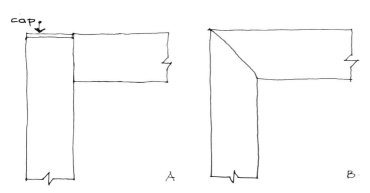

2.63 *A gate frame made up of steel angle or rectangular hollow section is best mitred (B) rather than butt jointed (A). If butt jointed a rectangular hollow section would need to have caps welded over its open ends.*

221

sections, the corners of the frames would have to be either butt welded or (better) mitred before welding (Figure 2.63).

Alternative ways of giving a welded metal gate curvilinear qualities, is to use strip material, or to cut the pattern from a metal sheet. Figure 2.64 illustrates a design where rigidity has been given to the gate through an overall pattern into which metal strip could be (relatively easily) bent, and another (Figure 2.65) where the pattern has been incorporated within a stressed skin approach.

2.64 *Mild steel strip offers many opportunities to the gate designer. In the example shown here the necessity for bracing has generated a pattern.*

2.65 *As noted in the caption to Fig. 2.59, a gate need not have a brace if distortion can be avoided in other ways. In this example the pattern in the neighbouring wrought steel fence is replicated, in the negative, by cutting into a mild steel sheet.*

The following list gives steel/sections which might be useful when designing metal gates.

1. Bars. Circular and square sectioned steel bars are made with diameters and side lengths in millimetre increments starting at 10 mm (⅜″). However, not all of these increments are regularly made and using dimensions which are a multiple of 5 mm (³⁄₁₆″) is likely to make finding a source of supply less difficult.
2. Flats. Solid rectangular sections are made with widths which increase in multiples of 5 mm (³⁄₁₆″) and with thickness in millimetre increments starting at 1 mm. The very thinnest of these are unlikely to be of use in gate design for they would have very little strength; and again all of these sizes are not commonly manufactured. A 5 × 10 mm (³⁄₁₆ × ⅜″) flat might be considered as the smallest useful size for this purpose and, again, specifiying in terms of 5 mm (³⁄₁₆″) increments should find suppliers.
3. Hollow sections. Hollow steel sections (circular, square and rectangular) are available in a range of sizes, and most of these

222

also have alternative wall thicknesses. A section with a thicker wall will be stronger than alternatives of the same overall size. This can be particulalry useful when a design calls for sections all of the same size, but when some of those will be more highly stressed (those to the basic frame, for example). Some sections are tabulated below:

Wall thickness mm (in)

CIRCULAR HOLLOW SECTIONS

Diameter mm (in)	2.5 (0.1)	3.2 (0.13)	4.0 (0.16)	5.0 (0.20)
21.3 (0.84)		x		
26.9 (1.06)		x		
33.7 (1.37)	x	x	x	
42.4 (1.67)	x	x	x	
48.3 (1.9)		x	x	x
60.3 (2.37)		x	x	x
76.1 (3)		x	x	x

SQUARE HOLLOW SECTIONS

Side length mm (in)	2.0 (0.08)	2.5 (0.1)	3.0 (0.12)	3.2 (0.13)	4.0 (0.16)
20 (0.08)	x	x			
25 (1.0)	x	x		x	
30 (1.18)		x	x	x	
40 (1.57)		x	x	x	x
50 (1.97)		x	x	x	x
60 (2.36)			x	x	x

RECTANGULAR HOLLOW SECTIONS

Side length mm (in)	2.5 (0.1)	3.0 (0.12)	3.2 (0.13)	4.0 (0.16)
25 × 50 (1.0 × 2.0)	x	x	x	
30 × 50 (1.2 × 2.0)	x	x	x	
40 × 60 (1.6 × 2.4)	x	x	x	x

2.9.5 Gate-posts

Gate-posts can make or mar a design; good fencing with good gates can be wrecked by the posts. There would appear to be a general tendency to make the gate-posts higher than the fence posts, but this only adds to the importance of the opening without giving it the visual (and actual) strength which it needs. Stouter posts are needed structurally (for the whole of the weight of the gate will be cantilevered off of the hinge-post) and this also seems to give just the right degree of visual emphasis needed without the risk of over-statement. But, as so often is the case, the physical strength needed by the post will almost always be less than its necessary visual strength. It is most unlikely that a post which is visually strong enough, will not also be structurally sound (given, of course, that the materials are sound).

An argument was presented in section 2.6.3 which suggested that rectangular rather than square fence posts were generally advisable, and that argument went on to show that (with some exceptions) it is best for the longer axis of the posts to be set perpendicular to the general run of the fence. Sized and set in this way, the fence posts are best able to resist wind and other loadings. But the situation at a gateway is different. When shut, the weight of the gate will attempt to bend the post in the same direction as the general run of the fence, but when it is open (or opening) the stresses will be similar to those induced in the fence as a whole. Gate-posts are, then, best when their section is symmetrical about both axes, that is either square or circular.

(i)   Timber gate-posts. 100 × 100 mm (4 × 4″) square, or 125 mm (5″) diameter round timber gate-posts should be sufficient for a low narrow gate, say 900 × 900 mm (3 × 3′). With such a relatively light gate, the additional strength given by hardwood would not be a necessity, although that material might be preferred for other reasons.

With a taller gate of similar width, hardwood might be used of similar dimensions as given above, whilst softwood posts would be better if a little larger, say 125 × 125 mm (5 × 5″) square or 150 mm (6″) diameter.

150 × 150 mm (6 × 6″) square or 175 mm (7″) diameter should be satisfactory for each leaf of a double gate 900 mm (3′) or so high and up to 2.4 m (8′) wide overall (that is, with each of two leaves being up to 1.2 m (4′) wide) in either hardwood or softwood.

A minimum 175 × 175 mm (7 × 7″) square or 200 mm (8″) diameter round post in hardwood would be needed for a 900 mm (3′) or so high single gate up to 3 m (10′) wide, or 225 × 225 mm (9 × 9″) if in softwood.

Taller wide gates are likely to be too heavy for timber posts of dimensions that could be easily obtained and so steel might have to be considered.

It is even more important that gate-posts are firmly set than it is for the fence posts. If a gate post should move at all, it is likely that the gate will become difficult to open and/or the latch cease to work. It is not, then, recommended that gate-posts be set simply by driving. The notes in section 2.6.3 on setting timber posts in concrete apply here also.

It is also even more important that a gate-post does not deteriorate – and for the same reasons. Tops must always be weathered and it is best if they are given a metal or some other form of cap. See also sections 2.6.2 and 2.6.4 which discuss preserving, painting and staining timber.

(ii)   Metal gate-posts. As we said above, most gate-posts which look to be in proportion with their gates, are far stronger than is strictly necessary. This is particularly the case with metal posts. Indeed their

potential strength can be exploited to give very slim and elegant posts, which might complement or contrast with the neighbouring fence and or gates.

A 60 × 60 mm (2⅜ × 2⅜″) rolled hollow steel section, or 60 mm (2⅜″) diameter tubular steel section, would take the weight of a 900 mm (3′) high, light gate up to 3 m (10′) wide, whilst 70 × 70 mm (2¾ × 2¾″) square or 76 mm (3″) diameter hollow sections would be sufficient for a heavy gate of similar width. (Note also that some hollow sections are available with a number of different wall thicknesses. If a relatively heavy gate was to be supported, but it was

---

Wall thickness mm (in)

CIRCULAR HOLLOW SECTIONS

Diameter mm (in)	3.2 (0.13)	3.6 (0.14)	4.0 (0.16)	5.0 (0.20)	6.3 (0.25)
60 (2.36)	x		x	x	
76 (3.0)	x		x	x	
89 (3.5)	x		x	x	
114 (4.5)		x		x	x

SQUARE HOLLOW SECTIONS

Side length mm (in)	3.0 (0.12)	3.2 (0.13)	3.6 (0.14)	4.0 (0.16)	5.0 (0.20)	6.3 (0.25)	8.0 (0.32)	10.0 (0.39)
60 (2.36)	x	x		x				
70 (2.75)	x		x					
80 (3.15)	x		x			x		
90 (3.54)			x			x		
100 (3.94)				x	x	x	x	x
120 (4.72)					x	x	x	x

RECTANGULAR HOLLOW SECTIONS

Side lengths mm (in)	4.0 (0.16)	5.0 (0.20)	6.3 (0.25)	8.0 (0.32)	10.0 (0.39)
100 × 60 (3.9 × 2.4)	x	x	x		
120 × 80 (4.7 × 3.2)		x	x	x	x
150 × 100 (5.9 × 3.9)		x	x	x	x

ROLLED STEEL I SECTIONS

Length × breadth mm (in)

76 × 76 (3 × 3)	89 × 89 (3.5 × 3.5)	102 × 102 (4 × 4)	114 × 114 (4.5 × 4.5)	
127 × 76 (5 × 3)	127 × 114 (5 × 4.5)	152 × 76 (6 × 3)	152 × 89 (6 × 3.5)	152 × 127 (6 × 5)

ROLLED STEEL ⊏ SECTIONS

Length × breadth mm (in)

76 × 38 (3 × 1.5)	102 × 51 (4 × 2)	127 × 64 (5 × 2.5)
152 × 76 (6 × 3)	152 × 89 (6 × 3.5)	

wanted to keep the gate-posts as slim as possible, then a section with a thicker wall might be used.)

For a very heavy, tall, or wide gate, rolled steel I or Ϲ sections could be considered. However, designers would be best advised to seek the opinion of a structural engineer as to the minimum sizes in any such circumstance. Listed below are steel sections which might be found useful for this purpose.

Clearly, metal gate-posts could be fabricated from sheet and/or strip material, or might be cast or wrought, all to integrate within a one-off gateway design. Given that these were properly designed and constructed within themselves so as not to buckle under the weight of the gate(s), then essentially solid designs should be within the range of sizes described above whilst posts which incorporated tracery or lattices could be rather larger in overall size to both visual and structural advantage.

Some manufacturers of metal railings and fences also offer 'cast-iron' posts both for use within the general run of the fence and with slight modifications, for use as gate-posts. These manufacturers' technical literature should be consulted.

As metal gate-posts are most likely to be considered for use with relatively heavy gates, it is critical that they are very well set. A 450 or 600 mm (18 or 24″) diameter hole, 900 mm (36″) deep is likely to be necessary, with the post set in concrete that comes as close to ground level as the proposed ground treatment will allow.

Finally, reference should also be made to the notes on finishing metal posts and gates in section 2.7.3

2.9.6 Ironmongery

Ironmongery is the collective term used for all hinges, latches and similar metal items that are parts of a gate design. Again we have something of a misnomer, for rarely is modern ironmongery made of iron. The vast majority is steel. Almost all is prefabricated and selected by the designer from the manufacturer's or merchant's catalogue. Care must always be taken with design by selection. It might prove tempting simply to satisfy function without consideration of the visual effects of alternatives. Whilst essentially utilitarian, ironmongery also contributes to the character of the whole. Inappropriate selection can radically alter overall appearance. Inappropriateness can be expressed both through disparate character and lack of visual strength. This section considers hinges and fasteners, and in the process touches on some other items of ironmongery.

(i) Hinges. Basically, a hinge consists of two leaves of metal fixed together in such a way that whilst they can rotate relative to each other, they cannot pull apart. These leaves are fashioned so that one can be fixed to the gate-post and the other to the gate. With an

appropriate hinge near to the top of the gate, and another at its heel, a gate can open and close without difficulty. (Incidentally, hinges are conventionally counted in pairs. So, one does not usually specify that a gate is to have two hinges, but one pair. Somewhat oddly, this continues when a heavy gate has to have more than one pair of hinges – a gate with three hinges, for example, is said to have a pair and a half.)

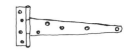

Tee hinges. Amongst the commonest form of hinge used with gates is the Tee hinge – so called because elevationally it looks like an emaciated capital T. Made of steel, their length varies between 254 and 610 mm (10–24″), in 51 mm (2″) increments. They are available in medium, strong, and weighty strengths, and either without any finish, or sheradized, galvanized, or black japanned. (Japanning is a lacquer-like finish applied by dipping.)

Steel tee hinges are most likely to be used with relatively narrow and lightweight gates. A length should be selected for the top hinge which is between one-third and one-half of the width of the gate, for it is this hinge which has to transfer the bulk of the weight of the gate, to the post. The heel hinge can be shorter, for the forces here are pushing into, rather than pulling away from, the post.

Japanned hinges need no further finish on site, but they will start to rust in a fairly short period. Whilst initially more expensive, it is better to use zinc-treated hinges, painted on site.

The heavier tee hinges are drilled to take countersunk screws, whilst round-heads are needed for those which are lighter.

The horizontal element of a tee hinge is screwed to the back of the rails of the gate, whilst the vertical element is fixed to the back face of the post (Figure 2.66). As this figures shows, a gate hung on tee hinges will always be recessed relative to the front face of the post.

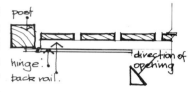

2.66 *T hinges are screwed to the back of the post and gate rails. It should be noted that this means that the gate can only open in one direction.*

Hook and band hinges. These are similar in form to a tee hinge but differ in one important respect. A tee hinge, once made, has its two leaves permanently attached to a common rod. A hook and band hinge has a loop formed in the end of the band (which is fixed to the gate) and this loop is placed over the hook (which is attached to the post). This is illustrated by Figure 2.67. This form has been inherited from the time when all hinges had to be hand forged rather than machine made. It is retained because of its strength. An additional feature is that the two leaves of the hinge are not permanently attached to each other, and so the gate can be lifted off its post without having to take out any screws or bolts. This might be an advantage or disadvantage, depending upon circumstances. If a disadvantage, a steel cap can be welded to the top of the hook after the gate has been hung (avoiding, of course, interfering with the operation of the hinge).

Hook and band hinges are commonly available up to 914 mm (36″) long. The smaller ones are from 305 to 457 mm (12–18″) long in 51 mm (2″) increments. The longer are 610 and 914 mm (24 and 36″) long.

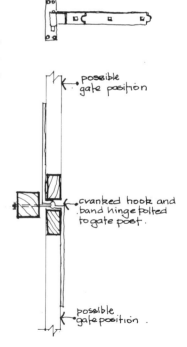

2.67 *A cranked hook and band hinge, when fixed to the face of the post, will allow the gate to open in both directions.*

227

Again, a length should be used that is between one-third and one-half of the overall width of the gate.

These hinges can be obtained without a finish, or galvanized. The hook (or pintel as it is sometimes called) may be welded or forged to a drilled and countersunk plate, for screwing to the post. Alternatively it may be an integral part of a bolt, with washers and a nut, which is then fixed by drilling right through the post (Figure 2.66). This figure also shows that when fixed in this way, a cranked band must be used if the gate is to be opened through 180°.

The band may be drilled and countersunk for fixing with screws, or simply drilled for bolting to the gate rail.

Double-strap hinges. These are made for very heavy timber gates. They are similar in principle to a hook and band hinge but, as the name suggests, have a double band. This is fixed with bolts running right through both bands and the rail. Such a hinge is illustrated by Figure 2.68. This also shows a short heel hinge with a spike fixing. As said above, the forces on the heel-hinge are into the post. It can, then, be both shorter and without a bolt. (Note that this spike is significantly more substantial than a nail. It is best fixed by driving into a drilling that gives a tight fit. Otherwise there is a real risk that the post will be split.)

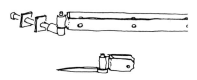

2.68 *Double strap hinges are necessary with very heavy gates, but the heel hinge can be relatively short.*

The hook of a double-strap hinge must be offset from the face of the post by at least half the thickness of the gate. If not the gate could not be fully opened (when hung between the posts) or shut (when hung on the post face).

Double-strap hinges are commonly available at 610 and 914 mm (24 and 36″) long and 51 mm (2″) wide and either with a galvanized finish or without. Short heel hinges are commonly 127 mm (5″) long. Both are made with a gap between the straps for a 76 mm (3″) thick gate, and with bolts for both 178 and 229 (7 and 9″) posts. However, such hinges as these can reasonably easily be made to order to fit gates and/or posts of other thicknesses.

All the above hinges have been discussed in terms of hanging timber gates to timber posts. The notes which follow consider other combinations.

Timber gates in masonry walls. Gates (of any material) cannot be hung directly on to a dry-stone wall, for the weight of the gate would pull stones out of the wall. The designer has three options. First, a timber post can be used and the gate hung as described above. Secondly, a large freestone post might be used (if the local stone produces such monoliths). The hinge hooks can then be obtained with long bolts taken right through the stone post, with any tolerances taken up with mortar. Thirdly, the dry-stone wall might be finished with a mortared pier at the gateway. The gate can then be hung as described below.

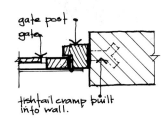

2.69 *Gates in masonry walls can be hung from timber posts fixed with fishtail cramps built into the masonry.*

The designer has two options when hanging a gate on a brick, mortared stone, *in situ* or precast concrete, or concrete block wall.

228

First, a timber jamb might be fixed to the wall by building-in with fishtailed cramps as shown in Figure 2.69. Tee hinges can then be screwed to the jamb. Secondly, the pintle to a hook and band hinge might be obtained that itself is welded or forged to a fishtailed strap. This can, then, be directly built into the wall. The heel hinge needs only a short, single fishtail, but the upper hinge would be more secure with a larger, double fishtail (Figure 2.70).

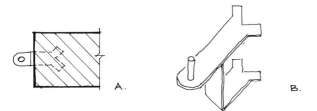

A.   B.

2.70 *Gates in masonry walls can be hung directly from pintles on fishtailed plates built into the wall (A). It might be advisable, particularly with heavy gates, for the upper hinge to be double fishtailed (B) with the vertical distance between the two tails being determined by the unit size of the masonry.*

Metal gates hung on metal posts. Proprietary hinges are not needed in these cases, for the hinge can be made as an integral part of the design. Pintles can be directly welded to the posts, for light gates, or forged to a plate which in turn is welded to the posts, when forces are greater. Tubular steel to take the pintles can be directly welded to the gate stile. This should have an internal diameter which is no more than a movable fit with the diameter of the pintle. Two pintles of 5 mm (0.2″) diameter and 50 mm (2″) long should be satisfactory for a low and narrow light metal gate, whilst 13 mm (½″) diameter, 75 mm (3″) long, at a maximum of 900 mm (36″) vertical centres should be satisfactory for all but the very heaviest of gates.

Metal gates in masonry walls. The gate should be designed as above with the pintles built in with fishtailed cramps as described earlier.

Timber gates to metal posts, and vice versa. It has to be said that these both would seem to be unlikely combinations due to the inevitably disparate characters. However, should such cases arise (and we would not suggest that that would be impossible) then the hook and band principle would be recommended, fixed to gate or post with screws, bolts or welds, as appropriate.

(ii) Fasteners. No gate could do its job if it were not fitted with some means of keeping it shut. Some must also have means of keeping them open. We will consider all such items under the general heading of fasteners.

Hooks. By far and away the simplest method of fastening a gate is with a cabin hook, and a typical item is illustrated by Figure 2.71. Cabin hooks are made in a range of metals, with steel being the most likely for use with gates. They are available in lengths from 100 to 300 mm (4–12″) in increments of 50 mm (2″). They can be obtained in self-finished malleable iron, which is rust resistant, or galvanized. Most have a plain shank but some are more decorative.

2.71 *Possibly the simplest way of fastening a gate, either open or closed, is with a cabin hook.*

Padlock bars. Cabin hooks will keep a gate shut but will not provide security. Padlock bars are similar in principle but are made of strip material. The bar (or hasp) once over the staple can be padlocked shut. The better items either have secret fixings or fold over themselves when padlocked and so cover the screws (Figure 2.72).

Padlock bars are made of hard steel with overall sizes in the region of 150 E 40 mm (6 × 1½″) or in heavyweight malleable iron at 250 × 63 mm (10 × 2½″) or thereabouts. These latter items are available self-finished, and both can be found galvanized.

Bolts. The principle disadvantage of cabin hooks for casual fastening, is that they leave the gate free to rattle. A bolt is more satisfactory in this respect. Figure 2.73 shows a bolt that can be padlocked either open or shut.

Bolts are available from 100 to 300 mm (4 to 12″) long with widths and metal thicknesses in proportion. The smaller items are generally black japanned whilst the larger are more normally galvanized.

All the bolts described above are for use where the throw of the shoot is horizontal. However, it can sometimes be useful to be able to bolt a gate down to a keeper held in the ground. This might be to avoid movement in a pair of wide gates when shut, or to hold a gate open when a cabin hook could not be used.

Vertical bolts have to have a relatively long throw, for they need to accommodate the gap below the gate. Figure 2.74 shows a typical vertical bolt which has a shoot of 16 mm (⅝″) diameter galvanized steel, is available in lengths of 305, 457 and 610 mm (12, 18 and 24″). Various lengths of throw can be accommodated as the tubes which hold the shoot captive to the gate, are in three sections rather than all fixed to the same plate. The keeper can be a simple plate screwed to a piece of buried timber but rather better is a detail where a pyramidal socket is cast into concrete where, because of its form, it will be securely held.

Again, versions of vertical bolts are available that can be padlocked.

Gate stops. Single gates can be restrained from swinging beyond the post (when this is wanted) either by cutting a suitably deep rebate into the post or, when it is not desirable to remove material from the post, by planting a timber section on to the post (Figure 2.75). But this is not possible with double gates. One of the gates must be restrained before the other can be made to close against it. This problem can be resolved with a gate stop.

The simplest form of gate stop is a short timber post that is, say, 50 mm (2″) taller than the clearance below the gate. The first leaf can then be cabin-hooked to the stop and the second leaf bolted to the first – possibly with rebates formed in the meeting stiles so that the pair are flush when shut.

However, this simple approach can be (to say the least) inconvenient when the gateway is to be used by vehicles which themselves have little clearance. There can always be the worry that the underside of

2.72 *A padlock bar offers more security than does a cabin hook but does not allow the gate to be casually fastened.*

2.73 *Some forms of bolts provide both a casual fastening and allow the gate to be padlocked.*

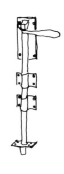

2.74 *A vertical long-throw bolt is needed at the meeting stiles of double gates if one of the pair is to be opened independently.*

2.75 *Gates which are intended to open in only one direction should have a stop either planted on to, or rebated out of, the post.*

the car might be damaged. The answer to this problem is to use a cast steel gate stop of the kind illustrated in Figure 2.76. This is, essentially, a box which can be cast into a concrete pad of about 250 × 250 × 150 mm (10 × 10 × 6″). The box has a hinged lid approximately 125 mm (5″) square which, when lifted, reveals a bolt keeper. The lifted plate stops the gate and allows it to be bolted. The more sophisticated versions of such a stop have a plate that will be caught and lifted by the bolt shoot when it is extended to an intermediate position, but will fall flat again when the gate is opened. However, paving levels must be constructed to fairly fine tolerances if this detail is to work properly.

2.76 *The meeting stiles of double gates might be stopped against a short post. However, this could damage the underside of vehicles or trip pedestrians. Cast steel gate stops are available with a counterbalanced hinged flap which falls flush with the ground when the gate is opened.*

Gate holders. The short post and cabin hook detail described above can also be used to hold a gate open when there is no adjacent fence or wall. A more convenient item is illustrated by Figure 2.77. These are most frequently made of self-finished malleable iron and the grooved stem cast into an *in situ* concrete pad, say 150 × 150 × 150–250 mm (6 × 6 × 6–10″). As the gate swings open the underside of the lower rail pushes the hook down whilst the counterweight ensures that it then engages. A flick with the toe allows the hook to disengage again. These gate holders are available with stems that are either 225 mm (9″) or 350 mm (14″) long.

2.77 *A gate holder, such as that shown here, will automatically catch and hold the underside of an open gate. A flick with the toe releases it.*

(iii) Latches. All but the last of the pieces of equipment described above need a positive action on the part of the user if the gate is to be held open or shut. Hooks have to be hooked or bolts bolted. Latches operate in such a way that catching the gate is automatic (given that they are properly fitted) and only releasing them again needs a deliberate action.

Spring catches. Agricultural gates are often fitted with a catch of the kind illustrated in Figure 2.78. The springy section of the catch is screwed to the inside face of the stile and the restraining double band screwed to the top rail. The large staple has a forged hook with a bevelled leading edge. This is driven into the inside face of the post (after starter drillings to avoid splitting) at a height that will catch immediately above the spring. The knobbed rod is long enough to give the leverage necessary to pull against the spring to release the catch. Spring catches are in wrought iron with 450 or 600 mm (18 or 24″) overall length. An alternative form of agriculture catch is shown in Figure 2.79. A flick with finger or thumb releases the catch. An advantage of this kind of latch is that it can also be padlocked.

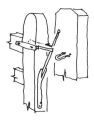

2.78 *Spring catches such as shown here were originally designed to allow a gate to be opened by a mounted rider.*

Gate latches. A very simple form of latch for small gates is shown in Figure 2.79. Such latches are made at 150 and 200 mm (6 and 8″) long in black japanned steel, and 250 mm (10″) long in wrought steel either galvanized or self-finished.

Latches of this kind (indeed most of the items described above) are fixed to the inside face of the gate, and so cannot be operated from the outside, except when the gate is low. Taller gates need one of two kinds of latches, both of which being developments of the simplest kind already described.

Figure 2.80 shows a latch where the metal strip is forged to a pivot rod. This rod passes through the gate stile and has knobs or rings both inside and outside the gate. Such latches are mostly of self-finished malleable iron and either 125 or 175 mm (5 or 7″) long. Cover plates can be variously shaped and decorated.

Whilst very tidy, perhaps even elegant, in appearance, these latches can sometimes need some strength to operate. If the pivot becomes at all rusty, or if the latch is too good a fit in the restraining hook, then a lot of leverage is needed to make it open. A thumb latch (Figure 2.81) is much better in this respect. Some also incorporate a pull handle, for use when the gate is to open out rather than in.

Lighter thumb latches, 150 or 200 mm (6 or 8″) long are made in black japanned steel, whilst heavier items 200 or 250 mm (8 or 10″) long, are available in galvanized steel. Other manufacturers produce variously wrought or forged iron items, and the less overtly antique of these can have a value when ironmongery with more character is desired. Simple latches of these kinds can also be made from hardwood.

The various fasteners described so far are specifically intended for use with timber gates. Some of the simpler hooks and bolts can be obtained without drillings for screw fixings, for welding to steel gates. However, none is technically complex and more satisfactory design is likely to result if the principles implicit in the descriptions given were adapted as appropriate for all-metal constructions. In particular, consideration might well be given to forming some of the parts of a bolt or latch directly in the fabric of the gate, rather than welding additional parts to it.

Finally, Figure 2.82 shows a detail of particular value with double gates – the large pivoted staple simply drops over the projecting stile of the other leaf.

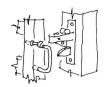

2.79 *The field gate catch (top) can also be padlocked, whilst the garden gate catch (below) cannot.*

2.80 *This latch, shown at the top, is pivoted on a spindle which passes through the gate, allowing the latch to be lifted from either side. The lower part of the figure shows a latch which can be useful when the design of the gate allows the latch to be reached either through or over it.*

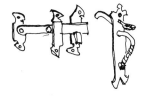

2.81 *A thumb latch provides a pull handle and allows the latch to be lifted from either side of the gate.*

2.82 *This item is really neither a catch nor a latch, but it does offer a very simple means of holding the meeting stiles of a pair of low gates.*

Details can invite further exploration both physically and psychologically.

Chapter three

RETAINING WALLS, RAMPS AND STEPS

3.1 INTRODUCTION

The proportion of the earth's land surface which is entirely flat and
level is very, very small – and almost all of that which is would be
quite impossible to build upon. Unfortunately, the proportion of
drawing boards which are flat is very, very high. It is all too easy to
design initially as if the site were flat and then introduce changes of
level into that design (as a somewhat belated operation) when the
nature of the site rather than that of the drawing board does finally
have to be confronted. This kind of sequential approach to designing
can never be fully successful. Whilst the scheme as built might be
technically satisfactory, even those concerns, most probably, could
have been better resolved had the existing topography been taken
into account from the outset. However, such matters are of little
significance when compared with the lost opportunities to introduce
incident and change of pace; to assist with the establishment of scale
and character; and to create and subdivide spaces with a range of
degrees of subtlety. This book is concerned about the design of space.
Space has three dimensions, and is experienced in four. If it is to be
effectively designed it must be conceived in four dimensions, not in
two, with the third and fourth grafted on at a later date.

Good landscape designers will provide themselves with a comprehensive set of existing levels from the beginning of the project. These
will include not only those within the site but also those beyond the
boundary. The distance these extend will vary with circumstances.
Clearly those immediately beyond the boundary are essential, for
physical fit between that which might be changed and that which will
not, must be achieved. But in some circumstances the relative levels
must be checked for some distance away from the site. The vantage
points from which the site might be seen, and those places which can
be seen from the site, must be identified – as must the ways in which
space flows up or down to, and through and beyond, the site.

The landscape designer must also check the intended finished floor
levels of any buildings which are yet to be constructed. (Indeed, it is
not unknown for architects to have to be reminded that their drawing
boards are also flat whilst the site is not.)

The strategic alternatives for the disposition of changes of level
must then be considered. These will be conditioned both by the

opportunities presented by the relationships between the site and building levels, and by the uses to which they are to be put. A site (or part of a site) which is to become a pedestrian thoroughfare would clearly need a radically different kind of treatment than one which should provide spaces in which people can pause to browse in a shop window, eat their lunch-time sandwiches, play, or simply sit and chat.

The key words are **space**, **place** and **pace**. Designers must think about the pace of the activity they are designing for. If it is essentially static then the space should be contained in such a way that it too is static. If it is active but self-contained then again the space should be similarly defined. If, however, the activity is one of strolling through a space, then the pace must be slowed – but without losing sight of the entrances and exits. Finally, as we work our way across this particular spectrum, if the requirement is to allow rapid and uninterrupted movement, then the space must be defined with clarity and directness.

Clearly, all of this does not apply solely to spatial definition through changes of level. The ways in which plants, walls and fences are disposed, and the visual and physical qualities of the materials used to pave a place, will all have a critical role. The point that is being made is that when considered as an integral part of the whole thinking, the form and location of changes of level will do much to amplify the qualities of the places created.

Only when the strategic objectives have been identified and broadly resolved, will there be a value in giving thought to tactical support. Not until it is known that a flight of steps, for example, needs to be broad and direct rather than narrow and intimate, will it be worthwhile to consider the materials and technical details of that flight. Not until decisions have been taken about the appropriate scale and character for a space can conclusions be reached as to whether it would be better to define that space with a vegetated earth bank, battered dry-stone retaining wall, bonded brickwork, dignified ashlar stone, or self-confident concrete.

Much of the sections which follow is best taken as an amplification of earlier parts of this book. It is not our intention to repeat, as examples, the basic technology for flagged paving or brick walls, but rather to show how the technology is varied when the flags are on the treads of a flight of steps or the brickwork is acting as a retaining wall.

The first section considers retaining structures, for these in some form or another occur at all changes of level. This is followed by a sub-section on ramps – essentially paving with a conspicuous slope. Finally, steps are considered for circumstances where a ramp with an acceptable gradient could not be constructed.

3.2   RETAINING STRUCTURES

If a granular material is loosely piled on a flat horizontal surface, then it will take a conical form. As more material is added to the top of the heap so its height will increase and its base broaden. But whatever the height and breadth of the cone, the angle between the sloping sides of the pile and the horizontal surface on which it stands will be constant – given, that is, that the material remains uniform. This angle is known as a material's angle of repose. It is the maximum angle at which that material could be banked and remain stable – provided that there are no subsequent changes in the condition of that bank.

Water is by far the most significant variable that a bank can experience. To illustrate through example, a clay subsoil will contain a fair quantity of water which binds the clay into a near homogeneous material. When first excavated a clay bank might appear to have an angle of repose of close to 90°. But the very act of excavation will allow ground water to start to drain out of the clay, and evaporate from its cut surface. As the clay dries, its capacity to support itself reduces and the bank crumbles away. Conversely, should there be heavy rain after the bank has been excavated, then the clay may become too wet to support itself. Large masses of the face of the near vertical bank will shear off in a distinctly more dramatic fashion than the slow crumbling as it dries.

Further, surface water run-off down the face of an unprotected bank will physically erode that surface, carrying away material to deposit it at the foot, when the rate of flow is reduced. Surface water erosion is a rather different concern to angle of repose, for an unprotected slope will be eroded to well below the angle at which it might otherwise have remained stable. (Clearly, the rate of erosion will also vary with different materials. Solid or broken rock would prove distinctly more resistant than would silts or sands.)

The simplest way to protect a bank which might otherwise erode is to establish vegetation upon it. But care must be taken to ensure that the vegetation will establish before erosion has proved to be a problem. Seeding grass just before a heavy rainstorm would not prove effective, whilst turfing the bank probably would. The maximum gradient of a grassed bank will be conditioned by the way in which the grass is to be cut, rather than concerns about angle of repose. Even a short slope at, say, 30°, would be hard work to cut with a hand machine and could be positively dangerous for a tractor-drawn mower.

This problem of maintenance can be avoided by protecting steeper slopes with woody plants. The branch and leaf cover breaks the heavy rain and, much more importantly, the root system binds and holds the surface. However, the difficulty remains of protecting the bank until the plants have grown sufficiently. This can be ameliorated by constructing a cut-off drain running parallel with the top of the bank. This will ensure that surface water run-off is limited to the rain that actually falls on the bank. However, this would not be a great deal of help unless the slope of the bank were comparatively short.

In extreme cases a herringbone arrangement of French drains built into the slope itself can avoid disaster; but this approach is moving beyond the scope of this book, being more applicable in association with engineering projects.

Another technique which is also not strictly within the scope of this book is to peg loosely-woven hurdles of willow or poplar species on to the surface of the bank. These need little encouragement to strike roots from cuttings and so, in most cases, a thicket of scrub develops to extend the initial protection offered by the hurdles themselves. Whilst this is an attractive notion, both philosophically and economically, it is somewhat crude as a planting design tactic. The number of appropriate species is rather small and the resulting scrub distinctly unkempt, making the range of contexts in which it might be applicable very narrow. However, this technique does allow an introduction to a group of others with wider possibilities.

3.2.1 Reinforced banks

Banks far steeper than the 35% to 45% which is the natural angle of repose of most subsoils can be constructed without initial problems of erosion, if the face of the bank is reinforced with hard rather than soft materials.

Gabions (rock-filled galvanized wire baskets) and ballast and cement-filled sandbags will not be discussed in detail here for, again, their appearance is such as to associate more with engineering rather than architectural projects.

(i) *Dry-stone walls.* Dry-stone walls built with a batter will effectively reinforce the face of a bank. Only a single skin of stone is needed as opposed to the two skins with rubble fill required for a free-standing wall (see section 2.5.3). Stone which gives rubble with near parallel faces gives a more effective mechanical bond, and that which gives large individual stones allows a thicker skin to be constructed. The better the bond and the thicker the skin, the taller and/or less battered can be the wall. Local observation can be important here, for the traditional details will have been developed and tested over centuries. In the absence of any other evidence it would be recommended that a limit of 15° batter from the vertical be put on the steepness of the wall, and structural engineering advice be sought if the height of the wall needed to exceed six times the thickness of the stone skin.

Where the stones are less regular, or the skin less thick, steepness and/or height should be reduced. At gradients of less than 45% this technique produces rip-rap – the limit of the potential of very irregular or spheroid stones.

The bank should be cut to the same gradient as the intended battered face, and a little further back than the width of the broadest stones to be used. Where bedrock is sufficiently close to the surface, the skin of stone can be built directly upon it. Elsewhere an *in situ* concrete foundation (section 2.1.1) 150–225 mm (6–9″) thick and 750–900 mm (30–36″) wide should be constructed, dependent upon the thickness and height of the stonework and load-bearing capacity of the substrate. A strip foundation might usefully be given a sloping top surface equivalent to the degree of batter to be given to the wall. The less steep the stonework the more important this could be, for in those circumstances there is an increased likelihood that the weight of the stone above would push out the foot of the battered wall.

The dry-stone skin is built upon the foundation, with variations in individual stone widths being taken up with consolidated selected excavated material bedded behind. The greater the irregularities in the back of the stonework, the better will be its key with the bank. Finally, the wall needs to be completed with an appropriate coping (section 2.1.5). Figure 3.1 shows a typical detail using relatively regular Cotswold stone.

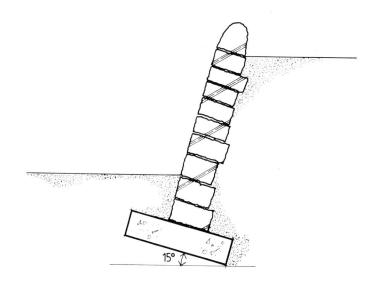

3.1 *A single skin of dry stone walling, with a 15° batter, could retain a height six times the thickness of the stone skin.*

With a rip-rap reinforced bank, it may be necessary to consider giving a foundation to the bank itself. The necessary strength of that foundation, will depend upon the magnitude of any loads which might be imposed upon the stonework (if, for example, a lorry might be tempted to run on to it). Section 1.15.1 is also relevant here.

(ii)   Crib walls. Crib walls offer a further variant on reinforced banks. A crib wall consists of a three-dimensional lattice of timber or precast concrete. The antecedents of this technique were relatively massive timber structures somewhat crudely jointed together in the manner of a log cabin. The interstices between the logs were filled with large rocks to give the whole structure sufficient mass to remain stable.

239

Commercially available precast concrete systems will allow the construction of equally massive crib walls, but with rather better structural predictability. Less substantial systems in concrete and timber, whilst having limits on their retained heights, give a smaller and more human scale. Manufacturers' instructions should always be most carefully followed and their representatives will offer specific advice if there is any doubt.

Clearly, concrete systems have greater longevity than do timber, but the treatments now available for timber (section 2.6.2) allow the psychologically softer qualities of that material to be exploited.

Prefabricated crib walls can have their interstices filled with selected as-dug material. Alternatively, where the excavated materials are heavy clays and/or silts, the crib walls can be filled with imported (or part imported/part as-dug) materials with better drainage characteristics. These qualities can all be further improved by including a land-drainage system behind the base of the crib wall. All this is to ensure that the fill is not washed out of the face of the structure. This is necessary not only to preserve the appearance and structural validity of the system but also to allow plants to be grown in the gaps between the horizontal elements at the face of the crib wall. These must be selected with some care, for they must be very tolerant to drought. But when successfully established they can, if it is wanted, completely disguise the true nature of the construction. It should be obvious that the shallower the slope given to the face of the crib wall, and the more open that face is, the larger and denser such planting could be. Figure 3.2 shows a typical crib wall system.

(iii) Perforated slabs. One more approach to reinforcing a soil bank can usefully be discussed. Perforated slabs in both precast and *in situ*

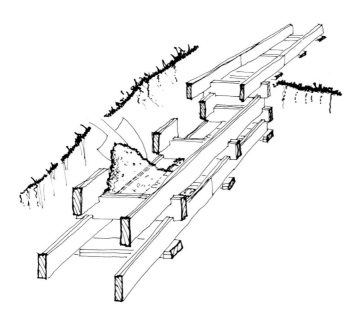

3.2 *Crib wall systems are available in both timber and precast concrete. These are given stability by the weight of material contained within their three-dimensional lattice. Vegetation can be encouraged to establish in the voids in their face.*

concrete are described as paving materials in section 1.15.7. These could be used on banks with slopes up to 45%, laid directly on the subsoil, when free-draining, and then filled with topsoil – in effect a form of rip-rap. (On steeper banks their stability would be suspect and a prefabricated crib wall system is likely to prove more effective.) These slabs will reinforce the bank against initial erosion whilst grasses or other ground-cover plants are established. If larger plants were wanted, some slabs would have to be omitted, the steepness of the bank would have to be limited and the remaining slabs laid to a broken and/or diagonal bond, all to ensure that the stability of the slabs above those omitted was not adversely affected. The inclusion of land drains at the top and bottom of such a reinforced bank should also be considered.

3.2.2 Retaining walls

A retaining wall is a vertical structure designed and constructed in such a way that the finished levels on either side of it are stable but at different heights. The angle of repose of a material was discussed in section 3.1. The work which a retaining wall has to do is to contain that wedge of material which occupies the space between the angle of repose and the vertical. Clearly, the taller the wall and the lower the angle of repose, the bigger that wedge will be and so the greater the strength needed to retain it. Equally clearly, a further variable is the relative weight of a unit volume of the retained material.

Analysis shows that a retaining wall must simultaneously perform in up to three different kinds of ways if it and the ground behind are to remain stable.

First, Figure 3.3 shows that the wedge of retained material can be considered as exerting a horizontal force on the back of the retaining wall. This force must be successfully resisted by ensuring that there is sufficient friction between the underside of the foundation to the

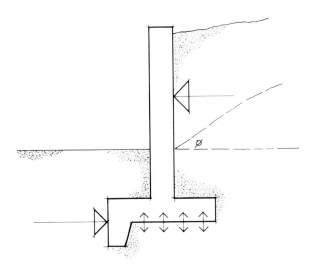

3.3 *The weight of material behind a retaining wall will attempt to push it forward. This must be resisted either by friction below the foundation and/or by the resistance offered by the subsoil at the face of the foundation.*

241

wall and the substrate below, and/or that it is so firmly fixed into the substrate that it is impossible for it to move horizontally.

Secondly, the retaining wall must not overturn (Figure 3.4). Over-turning can be resisted by giving the wall a mass such that (with gravity) it exerts a greater vertical force than the horizontal loads produced by the wedge of retained material. Alternatively, the wall must penetrate sufficiently deeply below the lower of the two finished levels that overturning is avoided by side resistances below the pivot point.

Thirdly, given that the retaining wall will not slide on its foundation, and will not overturn as a whole, then it must not break. It must be able to resist bending stresses and it must not shear (Figure 3.5).

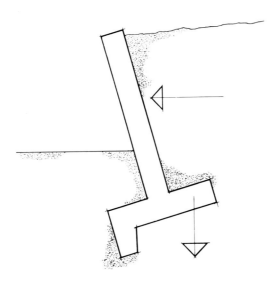

3.4 *A retaining wall that does not slide on its foundation may try to overturn. This must be resisted by the righting moments generated by the self-weight of the wall (possibly augmented by the weight of the retained material below the angle of repose).*

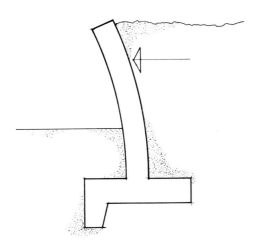

3.5 *A retaining wall that does not slide on its foundation and does not overturn will try to bend. The wall must be strong enough not to break.*

The nature of the different building materials of which a retaining wall might be made, and of their different associated technologies, allow different approaches to achieving stability. These individualities will be discussed in the sections which follow, but there remains one important general matter which should be considered first. The weight of wet subsoils will always be greater than when they are dry. If ground water were allowed to back up behind a retaining wall, then the loads which would have to be resisted would very significantly increase. It is generally uneconomic to design a retaining wall so that it has the strength to cope with wet retained material. So, in all of the sections which follow, ways and means of draining behind the retaining wall are also considered.

(i) *Timber*. Timber offers one of the technologically simplest ways of constructing a retaining wall. Individual timbers placed vertically and shoulder to shoulder, as shown in Figure 3.6, will satisfy structural functions in much the same way as do fence posts (section 2.6.3). The horizontal forces imposed on the upper part of the timbers by the weight of the retained material result in other horizontal forces imposed upon the subsoil by the lower part of the timbers as they attempt to pivot at or about the lower ground level. Provided that the side area of the timbers below the pivot point is large enough to spread those forces over a sufficient area of the substrate, then the wall will be stable – given, of course, that the timbers do not break through bending or shear. The retained height should never be greater than 8 to 10 times the thickness of the timbers and at least 40% of their overall length should be buried below the lower ground level. For example, a 150 mm (6″) thick timber retaining wall could comfortably retain a 1 200 mm (48″) high vertical bank provided that the timbers had an overall length of at least 2 000 mm (80″) – given that the foundations were properly constructed, that the retained

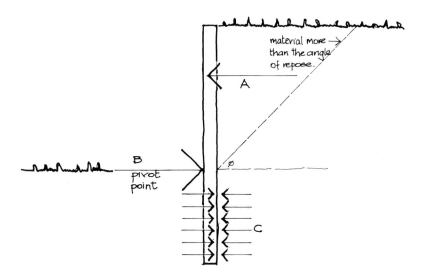

material more → than the angle of repose.

A

B
pivot
point

C

3.6   *A retaining wall made up of a continuous line of timber posts must be buried deep enough below lower ground level such that the forces generated at A can be resisted by the subsoil at C when the posts attempt to pivot at B.*

243

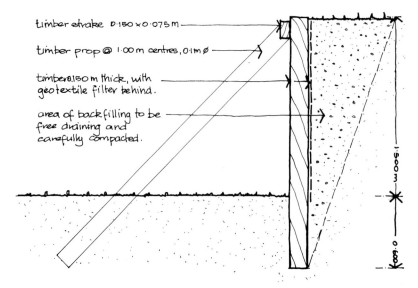

timber strake 0.150 × 0.075 m

timber prop @ 1.00 m centres, 0.1 m ⌀

timber 0.150 m thick, with geotextile filter behind.

area of backfilling to be free draining and carefully compacted.

3.7 *As an alternative to the detail illustrated in Fig. 3.6, the timber retaining wall might be given a propped strake.*

ground was properly drained, and that no significant loads (such as from vehicular traffic) were likely to be imposed on the upper level close to the retaining wall.

In theory, the loads to be resisted by each individual timber would be identical and so there would be no need to connect them to their neighbours. However, there are practical advantages in running a strake (horizontal member) close to the top of the wall on either the exposed or buried face, dependent upon visual concerns. Such a strake ensures that any locally imposed load can be spread over a number of timbers rather than being concentrated on a few, and also helps with alignment during construction. (But note that this would not be possible if the wall was to be freely curving on plan.)

The stability of a straked vertical timber retaining wall might be increased by using a prop to hold back the strake (waling) to the retained bank behind. Figure 3.7 shows one way of doing this. Such a detail greatly relieves both the bending and shear stresses in the vertical timbers, and the side pressures below the pivot point. But to be really effective the strake has to be on the exposed face of the timber wall. With such a detail 150 mm (6″) thick timber 2 100 mm (84″) long overall might be used to retain a 1 500 mm (60″) high vertical bank with no more than 600 mm (24″) buried below the pivot point.

A further variation replaces the bulk of the vertical timbers with horizontal sections. Figure 3.8 shows how overturning can be almost exclusively restrained through ties positively located in *in situ* concrete blocks. Each of the vertical timbers is tied, for the forces imposed upon the horizontal boards by the retained material must all be transferred to the ground through those verticals. That being so, they need to be rather more substantial than with an all vertical wall, say

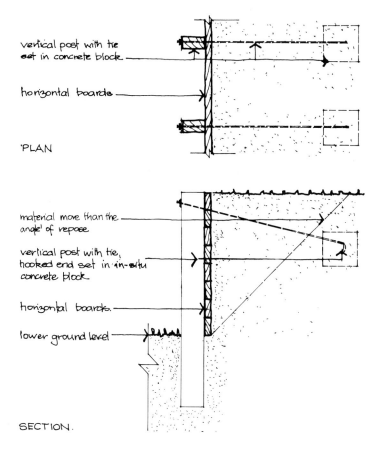

vertical post with tie set in concrete block

horizontal boards

'PLAN'

material more than the angle of repose

vertical post with tie, hooked end set in in-situ concrete block

horizontal boards

lower ground level

SECTION.

3.8 *A timber retaining wall might be detailed with boards between the retained material and a series of posts. The posts can be stabilized with metal ties hooked into* in situ *blocks.*

with a thickness of at least one-sixth of retained height, and at centres of ½ to ¾ of the retained height. Their width should be at least ⅓ to ½ of their thickness but this dimension is as likely to be conditioned by the need to allow a fixing for the horizontal boards. However, because such a wall is effectively tied, the penetration of the verticals need not exceed one-half of retained height (33% of overall length), provided that they are well set at that depth.

The width of the horizontal boards is of little consequence but their thickness should be at least one-twentieth of the centres of the verticals. It is critical that the horizontal boards are between the verticals and the retained material, for then the pressure can be transferred directly through the boards to the posts, rather than through any fixings. The advice of a structural engineer should be sought if such a retaining wall was being considered that would be over 1 500 mm (60″) high.

The posts to a timber retaining wall might be set in whichever of the ways described for fence posts in section 2.6.3 proves most satisfactory. Driving timbers shoulder to shoulder can be difficult for there is always the risk that as one is set it disturbs those adjacent. So, with an all vertical timber retaining wall, it is often most effective to excavate a trench centred on the line which the wall is to follow (say three times the width of the timbers to be used). The verticals can then be simply placed in position and temporarily propped whilst the trench is back-filled with compacted selected as-dug material, or *in situ* concrete.

The provision of drainage behind any timber retaining wall is not a conspicuous difficulty. Water will drain quite freely through the inevitable gaps between the timbers. This, however, can itself be a difficulty, for any fines in the retained material may be washed out. This would result in slumping at the higher ground level and, when paved, disfigurement of the lower. The answer is to back the retaining wall with an appropriate geotextile filter.

Back-filling can be with as-dug material when this is free-draining, or with graded granular material when it is not. Compaction should be undertaken with circumspection. Some compaction is necessary if the upper level is not to slump, but over-compaction could disturb the wall.

If unsawn timber is used, whole or split, it must be debarked for, with time, the bark would break away leaving rather larger than acceptable gaps between the timbers. All timber should be pressure impregnated with the best available specification for timber in contact with the ground (section 2.6.2). Sawn timbers could be stained and wrot timbers stained or painted (section 2.6.4). However, the authors have to confess to a bias against painted timber retaining walls for, in the main, the relatively sophisticated finish seems at odds with the essential character of the construction. But, as always, we remain open to the possibility that this bias can, sometimes, be proved unreasonable.

(ii) Masonry. Masonry retaining walls might be built of brick, concrete block, ashlar, or mortared rubble stone. All, in their simplest form, rely on their mass for stability. The horizontal forces from the retained material will attempt to tip the wall on its foundation, whilst the weight of the wall acts to keep it upright. The greater the overturning moment generated by the retained material, the greater must be the righting moment generated by self-weight. All this translates into a general principle that the higher the wall (thus increasing the volume of retained material) the thicker the wall must be (thus increasing its weight).

A useful rule of thumb for this kind of retaining wall, which includes a comfortingly wide margin of safety, suggests that they should be at least 200 mm (8″) thick, and at that minimum thickness the wall could have a retained height of 500 mm (20″). Thickness needs to be increased by 100 mm (4″) for every additional 500 mm (20″) of retained height. This rule of thumb can safely be used for walls with a retained height of up to 1 500 mm (5′) and, if there are no suspicions about load-bearing capacities, superimposed loads at upper ground level, or whatever, then it can be used up to 2 000 mm (6′8″).

246

It is not necessary for the wall to be of uniform thickness for the whole of its height. It can be thought of as a series of retaining walls built one above the other. Such a wall would be marginally stronger when it is the retaining face that is vertical and the stepped face that is exposed, for then the larger part of the weight will be acting vertically towards the rear of the wall (Figure 3.9). But this is not a necessity for there is more than adequate weight in such a wall for it to be the vertical face that is exposed if this is desired. (As an aside, a taller retaining wall can sometimes have an unfortunate effect on scale, and in such circumstances exposing the stepped face can help.)

Clearly the dimensions given by this rule of thumb must be adjusted to accommodate the unit dimensions of different materials. Further, allowances must be made within the thicknesses when mortar joints have to be introduced. Brick retaining walls, for example, would have a minimum thickness of 215 mm (8.5″) and would increase by width by 113 mm (4.5″) increments at 525 mm (21″) vertical intervals.

The foundations to an unreinforced masonry retaining wall are conditioned by exactly the same factors as are those for a free-standing wall (section 2.1.1).

A damp-proof course – other than one formed with two courses of engineering brick – should never be included in such retaining walls. As explained when discussing free-standing walls (section 2.1.2) a strip dpc will introduce a discontinuity. A masonry retaining wall would almost certainly either slide or overturn on such a dpc.

However, the saturation of masonry retaining walls can be a real problem. This is particularly the case when there is a risk of sulphate

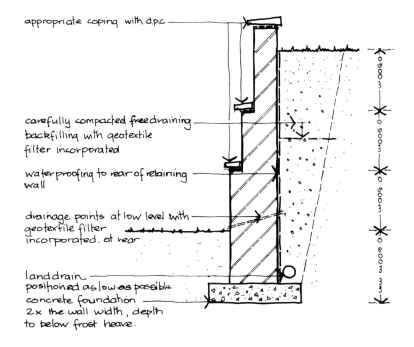

appropriate coping with d.p.c

carefully compacted freedraining backfilling with geotextile filter incorporated

waterproofing to rear of retaining wall

drainage points at low level with geotextile filter incorporated. at rear

land drain
positioned as low as possible
concrete foundation
2× the wall width, depth
to below frost heave.

3.9 *Masonry retaining walls rely on their own weight to achieve stability. Necessary weight reduces with height and so such a wall could be profiled as shown here.*

247

attack. A land drain should always be included immediately above the foundation behind the wall. Alternatively (or in addition in really wet circumstances) weepholes should be included by leaving perpends open at 1 000 to 1 200 mm (40″–48″) centres, 150 mm (6″) above the lower ground level. This, it has to be said, is a somewhat utilitarian detail that can be prone to becoming blocked and may dirty the lower ground level if it is paved. Both these conditions can be relieved by running a strip of geotextile filter behind the open perpends and lapped 100 mm (4″) or so above and below them.

If there remains a concern that the masonry may become saturated from behind, then a damp-proof membrane should be applied to its rear face. The simplest treatment would be two coats of bitumen or pitch-based paint, whilst a self-adhesive membrane or a 1:4 cement: sand render with added waterproofer, would offer greater security.

Readers are reminded that ensuring that the material behind a retaining wall is well drained is not solely a matter of protecting the wall itself. Water is very heavy. Should the retained material become saturated then the stresses on the wall will be very significantly increased. Ensuring that the retained material is well drained is, then, critical to the stability of the wall.

Expansion joints are a particular necessity in masonry retaining walls. The rear of the wall will always be relatively cool and stresses must be relieved at frequent intervals if the effects of differential movement are to be avoided. 10 mm (⅜″) wide straight and vertical joints must be left open, at 5 m (16′6″) centres. These must be protected from water seepage with a vertical dpc (Figure 3.10).

Masonry retaining walls would much prefer to be filled behind with sulphate-free granular material, graded, say, between 2 mm and 38 mm (0.1 to 1.5″). This ensures that ground water is not held in the fill but drains down to the weepholes and/or land drain. But this detail must be considered in the light of the way in which the ground above is to be treated. The fill cannot be effectively consolidated without risking disturbing the wall and/or adversely affecting the free-draining qualities of the fill. That being so, it is very likely to settle. Moreover, granular fill makes a less than ideal growing

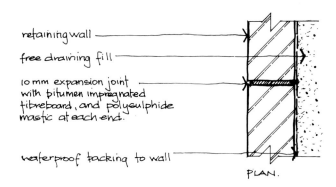

retaining wall

free draining fill

10 mm expansion joint with bitumen impregnated fibreboard, and polysulphide mastic at each end.

waterproof tacking to wall

PLAN.

3.10 *Long masonry retaining walls will move with a temperature variation and so must be given expansion joints at appropriate intervals, but these must be detailed in a way which does not allow the retained material (or water within it) to escape.*

medium for plant material. If the upper level is to be paved, then those pavings may have to be set on a reinforced concrete slab (section 1.13). Where the depth of fill is greater than the necessary depth of soil for planting, the granular material should be covered with a geotextile filter membrane at the appropriate level. If the depth of soil needed leaves no space for fill, then geotextile should be placed over weepholes and/or over the land drain, and covered with a minimum quantity of granular material to ensure that it is not displaced. If the available topsoil is heavy, then it would advisable to mix in a proportion of sand or grit (up to, say, 50%) so as to improve drainage.

Such retaining walls can, of course, be projected upwards to form a parapet. Indeed, the additional weight will add to the mass of the wall. But, at whatever level the wall is terminated, it must be given a satisfactory coping (section 2.1.5).

So far, this section has considered the design of masonry retaining walls when viewed in section. There can also be value in thinking about their plan form. The additional stability that can be given to free-standing walls by constructing them in panels or with buttresses, was discussed in section 2.1.3. Precisely the same principles apply to retaining walls, and again, such details offer wonderful opportunities to modify scale and introduce character into what, sometimes, might otherwise prove to be bleak and over-dominant features.

(iii) Reinforced retaining walls. It must be reiterated that the rule of thumb given for height and width at the beginning of this section applies only to masonry retaining walls of moderate height, and where ground bearing capacity below the foundations is good and there is no likelihood of poor drainage or superimposed loads at the upper level. Where this is not the case, the advice of a structural engineer must be sought. That advice is likely to include the use of steel reinforcement. That which follows does not set out how to undertake the necessary calculations. What it does do is explain the structural principles so that discussions with engineers can be more fruitful, and their advice better understood.

Retaining walls, reinforced with steel rods or mesh, can be considered to be of three different kinds.

Brickwork, blockwork and stonework can be bonded in such a way as to create voids within the wall. The form of construction will be essentially similar to that described in the previous section except that a number of reinforcing bars will be included in the foundation. Some of these will have a 200 to 300 mm (8″ to 12″) length left projecting upwards from the top of the found, positioned so as to coincide with the voids to be formed in the masonry above. Vertical bars are then wired to those starter bars. (This is for convenience – laying the *in situ* concrete around a forest of long bars would be very awkward, and would run the risk of disturbing the position of the bars.) The brick, block or stonework is then constructed around the bars with

249

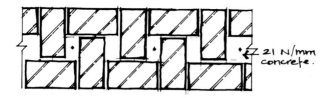

21 N/mm concrete.

3.11 *Quetta bond gives brickwork with vertical voids. These can be filled with reinforced concrete allowing a much longer, higher retaining wall than might otherwise be built with the same weight of masonry.*

the voids being filled with the strength of the concrete specified by the engineer, as the work progresses. Figure 3.11 illustrates Quetta bond in brickwork. The rule of thumb of 1:5 width:height given at the beginning of the last section, would give a maximum retained height of, say, 850 mm (34") for a 338 mm (13.5") thick brick wall. But reinforced in this way, a slenderness ratio of 1:8 might quite reasonably be achieved. So, an appropriately reinforced one-and-a-half brick thick Quetta bond retaining wall could have a retained height of 2.7 m (9').

Figure 3.12 shows how 450 × 225 × 225 mm (18 × 9 × 9") format hollow concrete blocks can be used so that the voids within the blocks themselves contain the vertical reinforcing rods and *in situ* concrete. Such a wall is potentially very strong simply because the voids are so emphatically contained. A slenderness ratio of 1:10 could be very reasonable – giving a potential retained height of 2.25 m (7'6") for a 225 mm (9") thick wall or 4.5 m (15') for a 450 mm (18") thick wall. However, a wall of such height would be somewhat tedious to construct, for the blocks have to be individually lifted and threaded

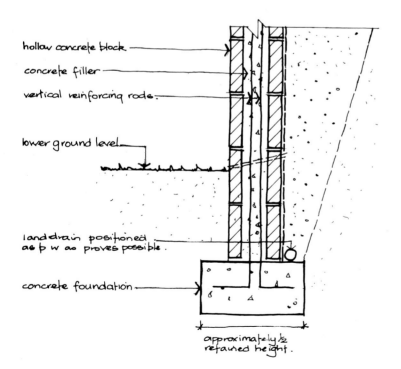

hollow concrete block

concrete filler

vertical reinforcing rods

lower ground level

land drain positioned as b w as proves possible.

concrete foundation

approximately ½ retained height.

3.12 *Hollow precast concrete blocks, with their voids filled with reinforced concrete, allow the construction of low but heavily loaded retaining walls. This detail is somewhat inconvenient for higher walls.*

250

over the projecting reinforcement. This difficulty can be reduced by using bars that are, say, the length of 3 courses plus a starter length necessary for wiring to the next higher bar; but lifting the blocks up and over the bars still remains hard work. So it is that this form of construction is most likely to prove useful with relatively low retaining walls but where stresses are high, resulting, perhaps, from having vehicles running close to the wall at the higher level. Such walls would all need to be considered in terms of damp-proofing, weepholes, land drainage, back-filling, etc. as discussed earlier.

Walls of the type described above give a series of reinforced concrete columns tied together with bonded masonry. If these columns need to be more strongly, and more predictably, tied together, then opportunities must be created that also allow horizontal reinforcement. The wall needs to be constructed as two leaves with a cavity. A vertical slab of reinforced concrete can then be constructed within that cavity as the work progresses. The masonry skins need not be very thick in themselves, for much of their structural purpose will have been served once the concrete has set. But, as discussed in section 2.3.4, it must be remembered that wet concrete is itself very heavy. The advice sought from the structural engineer should include the necessary thickness of the skins, the ways in which those skins should be tied together, and the maximum number of courses that can be constructed before allowing time for the concrete to set; in addition to the design of the reinforcement and specification for the concrete.

A wall of this kind could achieve a slenderness ratio of 1:12 without difficulty. However, if retaining over 2 m (6'8") the thickness of the outer skin may have to be increased, particularly near to the base, so as to buttress a relatively thin reinforced slab. Again, the principles of waterproofing, drainage, back-fill consolidation and the like would have to be satisfied.

Within a reinforced concrete retaining wall, maximum stresses occur at the point where the vertically cantilevered slab joins with the horizontal base, and so it is here that it needs to be thickest. In theory it needs to have no thickness at all at the top, but practicalities require 100–150 mm (4–6"). There can be real economies for walls over 2 m (6'8") high, particularly if over 25 m (80') long (the maximum length between expansion joints) in using no more than the minimum amount of concrete needed to withstand the stresses that will be experienced. But building the optimum section requires the use of temporary shuttering as described in section 2.3.4. Such a wall must be designed by a structural engineer; but prior to having full details to hand, designers can allow the thickest part of the vertical slab to be $\frac{1}{15}$ of the overall retained height, reducing to 100 to 150 mm (4–8") at the top. The whole range of finishes discussed in section 2.3.3 can be considered. Alternatively, the wall can be faced with brick or stone by casting wall ties into the *in situ* concrete at 900 mm (36") horizontal centres, so as to coincide with, say, every 5th course of the facing material. Again, drainage, back-fill consolidation and similar matters, need to be considered.

Reinforced concrete vertically cantilevered retaining walls are quite light relative to their retained height. Of all forms, these are most likely to slide on their base. They are, then, frequently designed with a down-stand toe to the base (as Figure 3.3).

An alternative approach projects the base back under the retained material. The mass of the retained material that is below the angle of repose assists the stability of the whole. This approach is frequently used by the very light precast concrete retaining systems that are available in lengths of 1.2 m (4') and heights up to 2.4 m (8') or thereabouts. Manufacturers' catalogues should be consulted, making particular note of the need to close the inevitable gaps between adjacent units. Whether *in situ* or precast, retaining walls of this kind require greater volumes of excavations and subsequent fill, but have a particular advantage when the design asks for plant material close to the wall at the lower level.

3.3   RAMPS AND STEPS

This section has been written assuming that the principles of paving construction are understood and the sections on the technologies associated with particular paving materials have been read and assimilated. Whilst cross-references to specific points are given, the generalities are not repeated.

This section considers the design and detailing of relatively short lengths of paving associated with distinct changes of level, be they banks or retaining walls.

3.3.1   Introduction

The broad relationship between a ramp or a flight of steps and the change of level of which it is to become a part can usefully be considered in terms of those qualities which are most apparent when viewed on plan, and also those which are clearer when considered in section. This is not to deny the arguments made at the beginning of this chapter, that landscapes must be designed holistically, but simply to say that analysis of the whole suggests these two parts. Both must be considered and integrated with each other and their broader context when designing. And further, designers must be conscious that whilst these qualities are most clearly displayed when viewed on plan or in section, those are not the ways in which they will generally be experienced. Designers must consider the ways in which (and the moods in which) people will move towards, up or down, through and away from, a distinct change of level. Views will be from those people's eye level – sequenced in both space and time. Moreover, the experiences will not be solely visual. The very sensation of bodily movement is of great significance, as are the sensed degrees of effort and changing pace.

Moreover, a flight of steps or a ramp will be a different experience for somebody moving up from a lower to higher level, than for somebody moving down. In the first case there are marvellous opportunities to arrive abruptly, without preview, at the higher space. An intimation might be given – by introducing formality into steps leading to a formal space, for example – or alternatively potential contrasts can be exploited to the full.

From the top of a flight or ramp, people can be given an overview of the space that they are about to descend into; introduced to the whole before they experience the parts, maybe even given the chance to decide which of the parts they choose to experience. Indeed, that overview will be inevitable unless the designer takes a deliberate decision to screen those views in some appropriate way.

Those qualities which are most apparent when the flight or ramp is viewed on plan can themselves be subdivided into two kinds. First, there is the nature of the line taken by the flight or ramp itself, and secondly there is the relationship between that line and the line of the bank or retaining wall that it is to cross.

A flight of steps or a ramp can take a very direct line, that is, follow a beeline between the upper and lower spaces. Such directness could be used to support a strategy of essential practicality, but equally might play a more positive role within compositions of simple geometries, be they with formality or informality. (It should be noted that this approach is not always possible. There are functional limits on the overall length of both steps and ramps which must not be exceeded if they are to remain safe. These are discussed later.)

The general line of steps or ramps might show angular changes of direction. Angularity, in this respect, breaks the flow of movement, and the more acute the angles the more abrupt will be that change of pace. Having to turn abruptly into and/or out of a flight or a ramp, even if the connection between is entirely straight, tends to make changing level an experience in itself, divorced to a greater or lesser degree from the places above or below. If the flight or ramp also makes angular changes of direction within its length, then that individuality can be heightened. This would particularly be the case if the landings that would be a practical inevitability at the changes of direction, were themselves given some identity – by exploiting a view, or providing a shade-giving tree, for example.

All of this contrasts radically with the potentials of a flight or ramp which curves within its length. Curves are essentially gentle, both visually and physically. A line of movement which curves into, through and out of a change of level, can allow total integration – given, of course, that such a line does not itself contrast radically with the character of the connected spaces.

The relationships between the general line of a flight or a ramp, and that of the retaining wall or bank which it is crossing, is also largely a matter of angle. A right-angled relationship will be self-assured and dominant, whilst the nearer the two lines come to being parallel, the more casual and self-effacing will the appearance be. All these points, of course, are made on the assumption that the detail expression of the steps or ramp does not contradict the effect of the broad relationships we have discussed. A casually gravelled curvilinear ramp with dry-stone abutments which projected at right angles from a straight and evenly graded grass bank, for example, would probably not look self-assured. It might, at best, look intriguing, but, equally, its appearance could be downright silly. Similarly, a flight of steps in ashlar, with rococo balusters to a corniced handrail, could never be self-effacing however narrow the angle between its general line and that of the bank or wall.

Those qualities of steps and ramps which are most clearly displayed when viewed in section, can also be considered to be of two broad kinds. The first of these is made apparent when the section of the flight or ramp is considered in itself, whilst the second is a matter of the spatial relationship between the steps or ramp, and the bank or retaining wall. The first of these points is essentially the same as the first of those made when steps and ramps were considered on plan. That is, that if there are no intermediate landings, then the greater will be the opportunities to integrate the experience with one or both of the connected spaces – and, conversely, the inclusion of landings (particularly when these are quite long relative to the length of flight or ramp leading to them) the greater will be the potentials to give that element individuality. But further, the angle of slope of a ramp, or the pitch of a flight of steps, will also have a bearing. Steep angles inevitably reduce pace. A steep flight such that makes people hesitate (even if unconsciously) before descending, or a steep ramp that requires bodily effort to climb (which would never be uncon-sciously) must make for a broken experience. Only the gentlest of slopes could be traversed with total unawareness of the change of circumstances, and only then if the ramp was quite short overall. That probably could never happen with a flight of steps.

Finally, but certainly not least importantly, a ramp or flight of steps might be built so that it projects forward from or is set into, the bank or retaining wall with which it is associated. Clearly, both of these relationships can be done with varying degrees of emphasis and, equally clearly, a ramp or flight could be of both kinds at different places within its length.

The experience of these different approaches will be radically dissimilar. The first projects the users out into space – detaches them from both the upper and lower levels. Those descending can be given a sense of lordliness consequent with their elevated position. Those ascending can be encouraged in their anticipations – the individuality of the element they are climbing saying that the place they are moving towards must, itself, be special. (But imagine the disappoint-ment if that does not prove to be the case.) And again, this can be

Sensitive detailing allows these slabs to appear to float upwards and onwards, signalling events yet to come.

done with a range of degrees of subtlety, from the visual equivalent of the classicism of a Mozart quartet to the monumentality of a Bach fugue.

But a flight or ramp that cuts back into the change of level encloses and contains its users. A curvilinear ramp softly rising (or falling) across a vegetated casual bank can induce a sense of secure well-being. In contrast, an angular and irregular flight of steps incising its way through a complex of vertical walls, will give an edge of mystery and uncertainty – a sublime rather than a beautiful experience; Byronic rather than Wordsworthian romanticism.

3.3.2 Ramps

(i) Gradients. Ramps are most frequently employed to allow wheeled vehicles to change levels, be they lorries, cars, perambulators, wheel-chairs, roller skates, trolleys or barrows. That being so, the important functional criterion that bears upon their design is their angle of slope. This is not simply a physical matter; a ramp which looks to be too steep to manage, is too steep – whether that is physically the case or not. So, a ramp connecting two different levels within a school playground could well be an entirely different design from one within the gardens of a retirement home.

Further, acceptable degrees of slope will also be conditioned by the slipperiness of the surface. And again, this will be both physical and psychological. A surface which is shiny, even if not physically slippery, could make for a frightening experience for the old or infirm. A surface which appeared to offer plenty of grip, but was actually only loosely bound, could leave a motor vehicle stranded or take a pedestrian suddenly by surprise.

Yet further, a ramp that is sometimes perfectly safe but at others very slippery, would be criminal; perhaps not literally so, but certainly could lead to a civil suit against the designer – and rightly so. As so often in landscape design, it is water on a ramp (in both its liquid and solid forms) that constitutes the greatest danger. Designers must ensure that surface water drains effectively off a ramp, both locally and generally. If it does not, then it is likely to be unsafe when wet and certain to be unsafe should that water freeze.

Having said all the above, we find it rather difficult to make categorical statements about acceptable maximum degrees of slope. So much depends upon the nature of the users, the surfacing materials, and the detailed design. And, an angle that might be acceptable over a short length, could well be unacceptable for a longer ramp. A vehicular ramp which drops to a junction with a road should be less steep at the junction than it might be elsewhere. A ramp which winds its way up a longish bank would need to be much less steep at its bends than within its lengths, otherwise the gradients at the inside of those bends will be very steep indeed. Designers must, then, treat any general advice with circumspection and be prepared to modify it in the light of the specifics of the circumstances for which they are designing.

As an example of the kind of general advice that is offered, one local authority suggests that the drive to a private house should ideally not have a gradient in excess of 10% (1 in 10) and that it should in no circumstances exceed 14% (1 in 7). 14% might be considered to be too steep for a vehicular access that must remain open in even the coldest weather, unless measures were taken to ensure that ice could not form on it (probably through heating). Also, 14% could look very steep to the infirm, or to somebody pushing a baby buggy or supermarket trolley. And 14% could well spell disaster for an invalid in a wheelchair or a toddler on a tricycle when descending.

(ii)   Surface water drainage. Allowing for surface water drainage off a ramp needs careful consideration, quite irrespective of concerns about slipperiness. Longitudinal gradients will be such that special efforts would be needed if water is to be made to drain towards gutters – and such efforts will only be fruitful with a limited number of paving materials. Moreover, water will run very rapidly down the ramp, whether in gutters or not. The probability is that a high proportion of the water would rush straight over any gulley gratings included in the length of the ramp, to accumulate in quantity at the landings or at the foot. Landings can, in fact, be useful here, Provided that gulleys can be located where they would not cause people to trip, catch pram wheels, look unsightly, or whatever, then landings can be used to break the general flow before it builds up into embarrassing quantities. If that does not prove possible, linear drains should be considered (section 1.17.4). These may be essential at the foot of a long ramp to avoid a large puddle forming. Linear drains may also be useful within the length of a very long ramp if its paved area is such as to collect more water than the drain at the foot could manage. The technical literature of the manufacturer of the selected drain will advise on necessary intervals.

(iii)   Abutments. The principles considered in the sections on Kerbs and Edgings (sections 1.17.2 and 1.17.3) apply equally here. But it must be remembered that in most cases the ramp's surface will be at a different level from that of the adjacent ground. Only when the ramp rises more or less straight up a bank that itself has a gradient of the order of 10% or less, could a relatively simple edging be used. More often than not, the ramp will either be above or below neighbouring ground levels – and not infrequently one side of a path will be in cut whilst the opposite is on fill (Figure 3.13). In such circumstances the kerbs will also be acting as retaining walls. Indeed, the edging to the ramp may have to be physically and visually contiguous with retaining structures. The points which have been made about retaining walls in section 3.2.2 must be reconsidered in the light of this additional role. Walls which are retaining the side of a ramp may

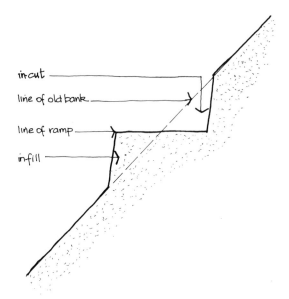

3.13 *A ramp running up and across the face of a bank will be partially on relatively stable cut and partially on relatively unstable fill.*

well have loads to resist in addition to those generated by the retained material. The wall might complicate land drainage. Foundations for the retaining walls might have to be taken much deeper than would otherwise have been the case to ensure that they are constructed on undisturbed ground. And, those foundations might have to be stepped as the wall climbs a bank.

(iv) Surfacings. A grass path (possibly reinforced with chippings) might be allowed simply to meander up an existing vegetated bank provided that it could be wholly cut into the bank and without the gradient on the uphill side of the path having to be cut at an angle that could not be stabilized. When this can be done there is no need to take precautions other, perhaps, than to include a cut-off drain to the rear to make certain that a stream does not form on the ramp during heavy rain.

Rolled gravel or stone chippings would be a satisfactory surfacing to a ramp which will have moderate pedestrian or very light vehicular use, provided that gradients are not too steep. Care must be taken that particle sizes are well graded (see section 1.11) and the paving and its hardcore base adequately consolidated. If not, the bulk of the surfacing would soon be at the foot of the ramp.

With steeper gradients and/or heavier use, tarmac, bitmac or asphalt paving would be needed so that the aggregates are bound in place. Untreated chippings should be rolled in. These will provide grip in wet and moderately frosty weather. Top-dressing with chippings and bitumen emulsion spray (section 1.7.8) will have the same effect provided that any loose chippings are brushed from the surface a day or so after the dressing is completed.

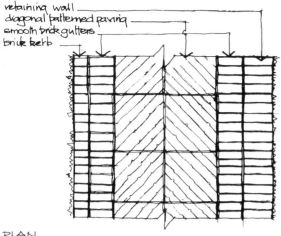

PLAN

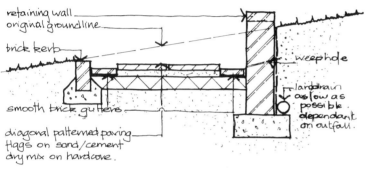

SECTION .

3.14 *Diagonally rilled flags can be used to pave the surface of a ramp. These will provide grip and encourage rain-water to drain to the sides of the ramp.*

But none of the surfacings considered so far, bound or unbound, will help to direct surface water into a gulley. Precast concrete flags can be obtained with various exposed aggregate finishes (section 1.15.3) which provide grip. Alternatively, some manufacturers make flags with diagonal rills which when laid to give an overall oblique or chevron pattern, greatly help in taking surface water to the sides of the ramp. Remember that smooth flags, brick pavers or similar will be needed in the gutters (Figure 3.14).

At extreme degrees of slope, clay or concrete bricks can be laid (on a cement:sand bed) using two different thicknesses of unit. This gives good grip and again, if laid to a diagonal bond, will conduct surface water to a gulley. Visual games can be played with details such as this by mixing colours and/or materials. But designers should be aware that the results of such games can be very brash – ideal in some circumstances, but less so in others.

All these flexible pavings can be laid in the ways described in the paving section, provided that a ramp has been wholly excavated into an existing bank. If the ramp is partially or wholly on fill, then either the ramp must be supported on the downhill side by a retaining wall,

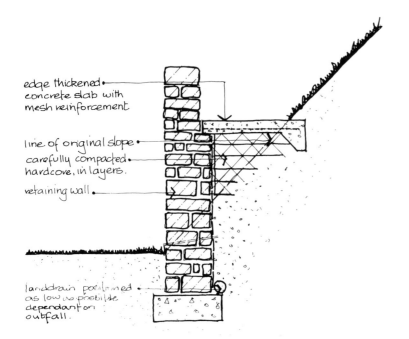

edge thickened
concrete slab with
mesh reinforcement

line of original slope
carefully compacted
hardcore, in layers.

retaining wall

landdrain positioned
as low as possible
dependant on
outfall.

3.15 *A ramp constructed on a considerable depth of fill might best be detailed with an in situ reinforced concrete slab.*

or a rigid form of paving used. Both of these would be advisable if a significant depth of fill was needed (Figure 3.15).

3.3.3 Steps

Few things have such potentialities for creating character as a flight of steps. For drama they are unbeatable. Think how Rome has exploited the levels of the Seven Hills. From the humblest flight connecting the alleys of Trastevere to that fantastic baroque staircase that flows down to the Piazza di Spagna, or the monumental ramped steps that lead up to the Capitol, each captures or creates (one cannot really tell which) the essence of the mood of the place. The Spanish steps are almost impossible to photograph in a still. They flow. The light moves over them like water. The steps to the Capitol are steady and measured and open. The alley steps are useful, shaded and almost secret. It is hard to imagine three similar things which are so utterly different.

There are countless other examples in less exalted places. The stone steps leading up to a loft, or down to a harbour wall, have their own monumentality; even an ordinary rung ladder has its temporary urgency communicating an airy sense of leading to somewhere unknown which will be gone tomorrow.

Steps might be divided into three main types, although like most categories they tend to overlap.

The solid, carved out (not necessarily literally), sculptural sort of step which is one with the rock or earth or wall to which it belongs.

The detached, flying, un-earthbound flights. These are often cantilevered. Or they may be little more than light ladders bridging the gap between levels. The more architectural types, which are really part of the building to which they belong, are usually a plinth to the building, e.g. the steps to St. Paul's, or an extension of an architectural idea, e.g. the steps to palaces like Versailles.

Each of these can in its turn have totally different character: leisurely or urgent, inviting or secretive, personal and private or social and open, monumentally imposing or casually informal. When designing a flight of steps, it must be decided which type best suits the site and materials available, but it is equally important to analyse the character of the site, so that the steps may reinforce it.

(i) Rise and go. The steepness of a flight will have more influence than anything else on its character. The first thing to note is that outdoor scale makes nonsense of the useful rule of thumb methods for calculating indoor staircases such as going × rise = 1 675 mm (66") or a twice riser + going = 585 mm (23"). In fact, steps based on these formulae seem to be amongst the least successful, particularly on wide flights. They appear mundane and purposeless. 280 × 150 mm (11 × 6") is stingy in the open. Steeper flights, which might be expected to be worse, have a character of their own. Either they smack of folly, or in certain positions they suggest privacy, perhaps indicating a back entrance as against a front, or a private flight off a public path. Shallow flights are generous and inviting, but a flight that could be taken two steps in a stride can become irritating.

As no satisfactory rule of thumb method seems to exist, the following examples of seemingly successful steps are given as illustrations. (All figures given represent a typical step – most flights vary slightly).

1. Athens; the steps up which one now climbs to the Acropolis (average) – 494 mm going × 175 mm rise (19½ × 7").
2. Rome; the Spanish Steps – 400 mm going × 150 mm rise (15¾" × 5⅞").
3. Rome; steps by Michaelangelo to Palazzo Senatorio – 400 mm going × 225 mm rise (15¾ × 6¾") and with these there is a 2–3% fall on the treads.
4. Rome; steps to the South East of Palazzo Conservatori – 370 mm going × 145 mm rise (14½" × 5¾").
5. London; St Mary-le-Strand – 343 mm going × 150 mm rise (14 × 6").
6. London; Kensington Palace Orangery – 300 mm going × 100 mm rise (12 × 4").

(ii) Ramped steps or perrons. These are much more common abroad (probably because wheeled traffic replaced pack animals earlier in Britain than in some of the remote hill towns on the Continent) but ramped steps are an excellent way of climbing long hills on foot. One can go steadily on, instead of taking the ascent in a shorter sharp

Flights of steps might be actually or apparently carved out of the ground or fly above it – to a radically different effect.

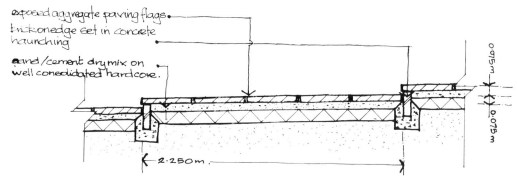

exposed aggregate paving flags.

brick on edge set in concrete
haunching

sand/cement dry mix on
well consolidated hardcore.

0.075 m

0.075 m

2·250 m.

3.16 *A perron might be
thought of as a ramp with steps
at intervals or as a flight of steps
with long slightly sloping treads.*

burst. Allow for an odd number, say three paces, on each tread. Even
if the slope on the tread should be shallow, say rising 75 mm (3″), an
overall gradient of 6.66% can be accommodated – far steeper than
could be the case with a ramp. Short flights of ramped steps with,
say, 50 mm (2″) risers can be negotiated by wheelbarrows, baby
buggies and the like, and so can be useful where a ramp would be
too steep but a conventional flight of steps impossible. Figure 3.16
shows a flagged perron.

(iii)   Single steps. Although generally deplored as dangerous, there
are countless examples of these which apparently cause no harm.
Every ordinary upstand kerb forms one. The first rule is that they
should be in an obvious position where they are unlikely to cause
surprise. An example of where they are unsafe is on an arbitrary line
which may topographically mean nothing, such as at a highway
boundary, or other change of ownership. But it does seem reasonable
to allow single steps where there is a change of function of a paved
area; particularly where a change of material can help to define the
step.

(iv)   Landings. Very long flights of steps without landings can
make people dizzy. There is also a danger that they will swing down
them at such a pace that it becomes difficult to stop. Breaks after
about a dozen steps should be considered and then a reasonably
generous landing of the width of a narrow flight or a couple of paces
on a wider flight. How generous the landing will feel will be very
dependent on the width of the flight. Local bye-laws should be
consulted. The Spanish steps in Rome (an example of a long flight)
never run for more than 12 steps without a pause. Landings vary
from approximately 1 to 2 m (3′3″–6′6″).

Concerns about the physical and psychological safety of flights of
steps are very similar to ramps. Long steep flights can be both
frightening and exhausting. Moreover, the consequences of a slip can
be much more severe. Breaking a flight with landings – ideally with
changes of direction, unless the landing is broad and the pitch

shallow – avoids apprehension and reduces effects should there be an accident.

And, again as with ramps, the landings on a flight of steps provide opportunities to deal with surface water drainage. It is not normally necessary to attempt to drain the treads themselves. Being flat and level individually, the effects of slipperiness are much less than with ramps. However, in areas of very high rainfall, or where rainwater cannot be taken off the flight until it reaches the foot, a narrow and shallow rill might be included at the back of the tread (where people will not trip over it). This need not be given a fall, for the amount of water in it will never be large. But a stepped gutter would have to be provided at the edges of the steps, and so handrails should be considered with care, so as to avoid people having to walk close to the gutter (Figure 3.17).

(v) *Treads.* In areas where frost can be anticipated treads are best constructed from textured materials. There can also be advantages if the treads are a lighter colour than the risers. The nosings will then visually contrast with their background, making them much more visible during gloomy periods.

(vi) *Foundations and edge restraint.* Finally, before considering different approaches to step construction in more detail, all of these

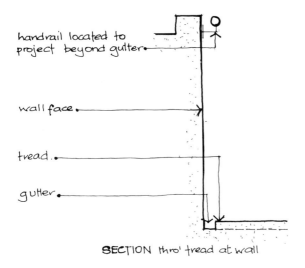

SECTION thro' tread at wall

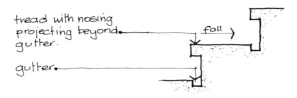

SECTION thro' tread + riser.

3.17 *In areas of heavy rainfall and/or hard frosts, the treads to a flight of steps might be given a slight backfall with tiny gutters which themselves fall to larger gutters at the edge of the flight.*

264

concerns about retaining the edges of the flight, and providing adequate foundation when on fill, considered in relation to ramps, apply here also.

(vii) Risers. A very casual stepped grass path is perfectly feasible – although if it were to have more than very infrequent use, the grass would have to be planted in a 50:50 chippings/topsoil mix. That being so, an identical technical approach could be taken for a flight with an all gravel surface, or chippings, bound with bitumen, tar, or asphalt. In our view, such a flight looks really well when it winds its way up a relatively gentle bank of irregular section. In such circumstances, and with a little thought, the steps can have all of the qualities of a naturally occurring feature – can appear to have grown into and with the evolving landscape. Further, such steps are much more satisfactory when cut into the bank about twice as deep as the riser height. The risers (Figure 3.18 shows these in split timber) can simply be held in place with a stout stake on the downhill side, near to each of their ends. This fixing will then disappear below the soil fill as levels are adjusted back to meet with original ground. The risers do not all have to be parallel nor the same length.

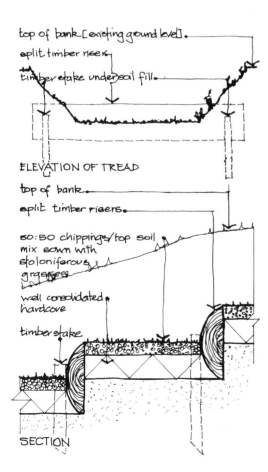

top of bank [existing ground level].
split timber riser.
timber stake under soil fill.

ELEVATION OF TREAD

top of bank.
split timber risers.
50:50 chippings/top soil mix sown with stoloniferous grasses.
well consolidated hardcore
timber stake.

SECTION

3.18 *Very simple informal flights of steps can be constructed with split logs pegged in place as risers, backfilled behind with hardcore and chippings.*

265

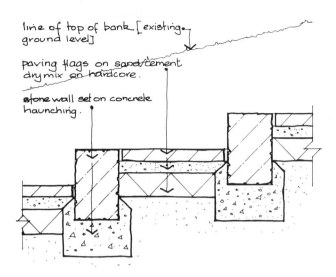

line of top of bank [existing ground level]

paving flags on sand/cement dry mix on hardcore.

stone wall set on concrete haunching.

3.19 *The principle illustrated by Fig. 3.18 is given more sophistication here using stone risers and flags.*

A very similar principle can be used for stone or precast concrete flagged steps. As Figure 3.19 illustrates, the risers would be constructed as a series of stone (or brick) walls retaining the hardcore and sand/cement bed for the flags. With such an approach it is important that the flight has a generous width. The abutting soil will spill on to the edges of each tread. This can be stabilized with ground-cover planting but, as should be obvious, that could not withstand trampling. If the flight is not over-wide for its amount of use, it will soon become very tatty in appearance.

(viii) *Abutments.* Where physical space is tight, making over-generous widths impossible, the treads cannot be allowed to die away into the neighbouring ground. Where the flight is cut into the bank, the neighbouring ground will have to be retained. Where the flight projects out from the bank, it will be the construction below the treads which has to be contained.

Where the first of these is the case, an early decision must be taken about the relationships between the top of the abutting (or abutment) wall, the line of the steps and original ground line. Is the top of the wall to follow the section of the bank, or the section of the steps, or an entirely independent line? If the former, there are relatively few materials which could be considered. Figure 3.20 shows a timber stake abutment which could be constructed to follow very closely the bank's section; a more or less independent line to that of the flight. A similar effect could be achieved in rubble stonework with a comber coping. But when using materials of regular unit size, this casual effect can only be achieved at a price, and it is debatable whether it would be worth it.

Regular units, such as bricks, are only going to be used economically when the top of the abutment wall also steps – but not necessarily synchronised with the steps in the flight. This immediately gives a visual emphasis to the flight – the rhythms of stepping become

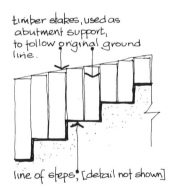

timber stakes, used as abutment support, to follow original ground line.

line of steps, [detail not shown]

3.20 *A timber post retaining wall as illustrated by Fig. 3.6 might be used as the abutment to a flight of steps. The top of such an abutment could be made precisely to follow neighbouring ground level.*

266

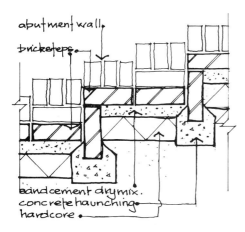

3.21 *An abutment wall built of regular units (such as the bricks in this example) cannot easily follow the profile of the adjoining bank. It must itself step, but these steps could set up a different rhythm to that established by the flight.*

dominant. Designers must accept and exploit this inevitability. Figure 3.21 shows a typical detail in brick.

(ix) In situ *concrete stepped slab.* The approach shown in Figure 3.21 is only economical when treads are quite wide. When tread widths are such that the individual strip foundations to risers would meet (or nearly meet) each other, a stepped concrete slab foundation would be much cheaper. Figure 3.22 shows a very straightforward approach, but one which does not have to be inelegant because of that. Basic construction is essentially similar to that described in section 1.13, but with, say, 150 × 50 mm, (6 × 2″) timbers used to shutter each riser. It might also be necessary to shutter the edges of the steps with 19 mm (¾″) plywood cut to profile. The concrete could have any of the finishes described in section 1.13.8. Indeed, there is no reason why treads and risers should not have different finishes. Or, treads and risers could be formed in flags or brick on the concrete slab.

Where the abutment walls are retaining little height, they might well be constructed on the same stepped slab, thickened at its edges to, say, 200 mm (8″) minimum. (But note that if this is to be done, then it would be best for the dimensions of the rise and go to co-ordinate with the unit sizes of the materials used for both treads and

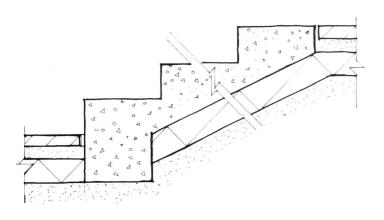

3.22 *A flight of steps might be constructed of in situ concrete laid on a hardcore bed. The risers are temporarily shuttered and a stiff mix of concrete used so that the weight of wet concrete above does not cause slumping to the treads below. Such a flight could be given a variety of finishes, or might simply be used as the structural base for flags or brick.*

267

the wall.) If the retained height is greater, or if the wall is carried up to support a handrail (and so is heavier) then it will have to be given its own stepped strip foundation as described for free-standing walls in section 2.1.1.

A concrete slab foundation will be essential for steps that are wholly or partially on fill. And unless there can be complete confidence that the fill can be thoroughly consolidated, it would be best to include a light mesh reinforcement in that slab.

(x) Permanently shuttered concrete steps. Steps which project forward of the bank or retaining wall with which they are associated, will almost certainly be on fill. A reinforced slab or series or slabs would, then, be most advisable. Figure 3.23 shows a simple detail. Concrete flags are set on edge on a concrete bed and haunch, and act as permanent shuttering for the supports to each tread. But this detail could only be used where the adjacent bank is only a little lower than

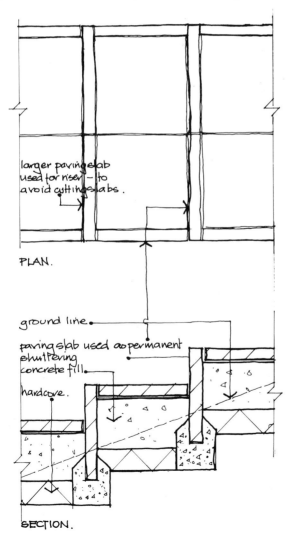

3.23 *Where a flight is to be built projecting forward and above the line of the adjacent bank, and without an abutment wall, consideration must be given to its appearance when seen from the side. In this example precast concrete paving slabs are used as permanent shuttering.*

268

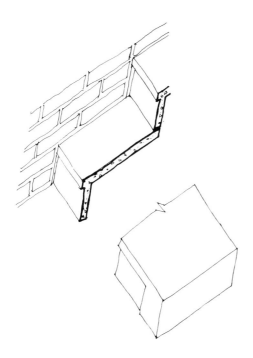

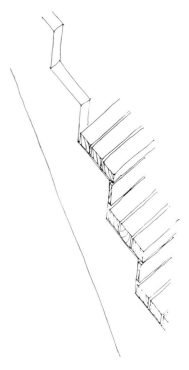

3.24 *Precast reinforced concrete can be used to make units incorporating both risers and treads. Such units would be valuable when a flight has to be constructed over poorly consolidated fill or over a void. In this example the units are given overall dimensions and a form which allows them to be easily built into brick abutments.*

the steps. The flags on edge cannot retain a great height for they would fail in tension if there were any suspicion of bending.

(xi)   Precast concrete. Where a flight wholly projects forward from the line of a bank (or, more likely, a retaining wall) reinforced precast concrete might be used for treads and risers cast as a single unit. The amount of reinforcement would be determined by the precast manufacturer's engineer so that the units were well able to support themselves and their loads over the necessary spans. The detail illustrated in Figure 3.24 could, with appropriate reinforcement, span at least 2 m (6'8"). That being so, there would be no need for any fill material below.

(xii)   Stairs and ladders. Flights of steps – which might better be described as stairs or, sometimes, ladders – can be designed to be supported on raking beams. These are known as strings when there is one at each end of the treads, or as spines when there is only one on the centre line of the flight above, with the treads cantilevered to each side. Figure 3.25 shows illustrative details in timber. Three relatively small sections are suggested for the treads, for economy, to avoid the risk of warping that there would be with a larger board, to provide some grip, and to give opportunities for rain to drain off the treads. The short riser simply reduces the gap which might trap the foot (or head) of a small child. This will also reduce any sense of insecurity that might be felt by the elderly, infirm, or timid. Sections of the sizes given would allow the stair to be 1 m (40") wide and 3 m (10') long with a comfortable factor of safety.

3.25 *This timber staircase has slatted treads screwed to a cut string. If, say 100 × 50 mm (4 × 2") timbers were used for the treads, then the strings could be at 1 m (36") centres. However, the flight overall could be as wide as was wanted provided that there were strings beneath the treads at those centres. Depending upon circumstances, it might be necessary to plant supports for a handrail on to the face of the string.*

Handrails should be 900 mm (36″) above the pitch line (a line joining the top edges of each nosing) with a radius top which is neither too fat nor too thin to grip. On longer flights it is advisable to design so that the hand will not slip off the end of the handrail in the event of a trip. Similarly, designs should be such that there are no large gaps between the handrail and string. This might be done with additional rails, with vertical balusters, or by introducing panels. Whilst these are all, essentially, safety features they will contribute a great deal to the character of the stair as a whole and, as always, this will be so whether their appearance has been carefully considered or not.

Fixing a timber stair at its head and foot will vary with the forms of construction at those locations. Clearly this must be very positive. Figure 3.26 shows one approach which is secure yet visually unintrusive.

Treating and finishing timber stairs with preservatives, paint or stain, will be as for timber decking (sections 2.6.2 and 2.6.4).

Timber might also be used for a stair with a spine beam as its major

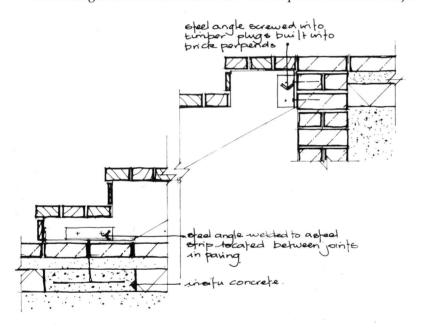

steel angle screwed into timber plugs built into brick perpends

steel angle welded to a steel strip located between joints in paving.

in situ concrete.

3.26 *It is as important that the way in which a stair is fixed at its head and foot is as carefully considered as is the design of the stair itself. In this example the foot is supported by a steel angle welded to a steel strip which is thin enough to be accommodated within the joints in the brick paving. The steel strip is split at its end and embedded in an in situ concrete pad. The top is located by steel angles screwed to timber plugs built into the perpends of the retaining wall behind. The fixings are, then, both substantial and invisible. However, because they are largely inaccessible, it would be best if the steel fixings were galvanized.*

structural element. But, if such is contemplated, the design should be evolved with a structural engineer. The cantilevered treads will put very considerable stresses on the joints between the treads and spine. Any movement at those joints will weaken the fixing, allowing more movement and further (and accelerating) weakening. Spine beams are best reserved for materials which more easily accommodate very positive fixings. Figure 3.27 shows such a stair in welded steel. Reinforced concrete has similar potentials (but different details). The chequered plate used for the treads is manufactured for precisely this purpose and is used, most frequently, in ships' engine rooms, and

3.27 *This stair, apart from its foundation pad, is built entirely from welded steel. The cantilevered spine beam and vertical support are of 200 × 100 mm (8 × 4″) rectangular hollow section, and so the support does not interrupt the pattern of the interlocking block paver below. Treads and risers, of chequered plate, cantilever from the spine beam. The handrail is a continuous strip welded to the treads.*

for gangways at oil refineries, power stations and the like. Whether there is some absolute quality in the material or whether it simply carries those associations, remains open for debate, but there can be no doubt that a stair such as we have illustrated has a strong engineering and machine aesthetic. Whilst this might be less strong with alternative details, and other associations might be introduced by using, say, cast and more overtly decorative elements, these kinds of qualities will remain. The contrast between the lightness and airiness of the whole, with the strength and solidity of the parts, gives tremendous potential to establish and reinforce the character of a space.

271

REFERENCES AND FURTHER READING

TAR, BITUMEN OR ASPHALT BOUND FLEXIBLE PAVEMENTS, AND COLD ASPHALT SURFACING

British Standards Institution

BS63: Part 1: 1987 and Part 2: 1987. Single sized roadstone and chippings.
BS76: 1986. Tars for road purposes.
BS434: 1984. Bitumen for road emulsions.
BS594: 1985. Rolled asphalt for roads and other paved areas.
BS1446: 1973. Mastic asphalt (natural rock asphalt fine aggregate) for roads and footways.
BS1447: 1973. Mastic asphalt (limestone fines aggregate) for roads and footways.
BS3690: 1982/83. Bitumens for building and civil engineering.
BS4987: 1973. Coated macadams for roads and other paved areas.
BS5273: 1985. Dense tar surfacing for roads and other paved areas.

Asphalt and Coated Macadam Association

What's in a road? 1985.
Construction and surfacing of parking areas for medium and heavyweight vehicles, 1985.
Decorative and coloured finishes for bituminous surfacings, 1986.
BACMI guide, 1986.

HMSO

Croney, D. (1977) The design and performance of road pavements.
Transport and Road Research Laboratory (TRRL) (1968) Protection of sub-grades and granular sub-bases.
TRRL (1981) Recommendations for road surface dressing.
TRRL (1966) Sources of white and coloured aggregates in Great Britain (Now out of print but copies may be found in libraries. Its content is covered in part by the BACMI guide.)

Other sources

Various technical publications are available from: The Mastic Asphalt Council and Employers' Federation (MACEF) and The Refined Bitumen Association (RBA); also from The Asphalt Institute, Maryland, USA.

CHIPPINGS SEALED WITH COLD BITUMINOUS EMULSION

Bitumen emulsions are sold under brand names so the manufacturers' instructions are the best references. Other information can be had from the various trade associations (Section 1.5).

SOIL–CEMENT BASES

Andrews, W.P., Soil-cement roads. The Cement and Concrete Association. Soil-cement roads. Fixed Equipment of the Farm, No.19, HMSO.

Specification for the construction of housing estate roads using soil-cement, HMSO.

Note. All the above were published in the 1950s and copies might be difficult to find.

IN SITU CONCRETE PAVEMENTS

British Standards Institution

BS12: 1978. Ordinary and rapid hardening Portland cement.
BS146: Part 2: 1973. Portland blastfurnace cement.
BS812: 1975/1985. Sampling and testing aggregates, sands and fillers.
BS882: 1983. Aggregates from natural sources for concrete.
BS1881: 1970/1984. Methods of testing concrete. (This has many parts and some have no application in concrete pavements.)
BS4550: 1970/1978. Methods of testing cement. (Again with many parts.)
BS3148: 1980. Methods of testing water for concrete.

Cement and Concrete Association

Shirley, D.E. (1985) *Introduction to concrete.*
Teychenne, D.C. *et al.* (1975) *Design of normal concrete mixes.*
Pink, A. (1978) *Winter Concreting.*
Shirley, D.E. (1980) *Concreting in hot weather.*
Shirley, D.E. (1984) *Sulphate resistance of concrete.*
Monks, W. (1984) *Appearance Matters; visual concrete.*
Monks, W. (1985) *Appearance matters; tooled concrete finishes.*
Monks, W. (1985) *Appearance matters; exposed aggregate concrete finishes.*

British Aggregate Construction Materials Industry

Anon. (1986) *The BACMI guide.*

INTERLOCKING BLOCK PAVEMENTS

British Standards Institution

BS6717: Part 1: 1986. Precast concrete paving blocks. (Covers manufacture of the blocks)
BS6717: Part 2 (in preparation). (Will cover construction of pavements using the blocks)

Concrete Block Paving Association (Interpave)

List of members and product guide, 1986.
Concrete block paving – Structural design of the pavement, 1986.
Concrete block paving – Surface design consideration, 1986.
Concrete block paving – Detailing, 1986.
Concrete block paving – Laying, 1986.

Cement and Concrete Association

Lilley, A.A. and Clark, A.J. (1980) *Concrete block paving for lightly trafficked roads and paved areas.*
Knapton, J. (1976) *The design of concrete block roads.*
Anon. (1980) *Specification for precast concrete paving blocks.*
Anon. (1983) *Code of practice for laying precast concrete block pavements.*

HMSO

A guide for the structural design of pavements for new roads, 1970.

The Brick Development Association

Publishes information specific to clay pavers.

SLAB, BRICK, SETT AND COBBLED PAVEMENTS

British Standards Institution

BS368: 1970. Precast concrete flags.
BS1217: 1986. Cast stone.
BS435: 1975. Dressed natural stone kerbs, channels, quadrants and setts.

Other sources.

Various publications are available from the Cement and Concrete Association, the Decorative Paving and Walling Association, the Concrete Block Paving Association (Interpave), the National Paving and Kerb Association, the Stone Federation, the British Precast Concrete Federation, the Concrete Brick Manufacturers' Association, the Brick Development Association, the National Federation of Clay Industries.
Ultimately, individual manufacturers' literature must be consulted and samples inspected.

TIMBER DECKING AND BLOCKS

See references under Timber fencing.

TRIM TO PAVED SURFACES

British Standards Institution

BS340: 1979. Precast concrete kerbs, channels, edgings and quadrants.

BS368: 1971. Precast concrete flags.
BS435: 1975. Dressed natural stone kerbs, channels, quadrants and setts.

Other sources

Various publications are available from trade associations (included in the list above) and manufacturers.

WALLS GENERALLY

British Standards Institution

BS743: 1970. Materials for damp-proof courses.
BS12: 1978. Ordinary and rapid-hardening Portland cement.
BS146: Part 2:1973. Portland blastfurnace cement.
BS4027: 1980. Sulphate-resisting Portland cement.
BS890: 1972. Building limes
BS1200: 1976. Sands for mortar.
BS5642: Part 2:1983. Coping units
BS1217: 1986. Cast stone.
BS4729: 1971. Shapes and dimensions of special bricks.
BS1449: Part 2:1983. Stainless steel strip.
BS2989: 1982. Zinc and Iron – Zinc alloy coated steel strip.
BS1470: 1987. Aluminium and aluminium alloy strip.
BS2870: 1980. Rolled copper and copper alloys strip.
BS1178: 1982. Milled lead sheet.

Other sources

Publication and advice are available from the various trade associations including Aluminium Federation, Brick Development Association, Cement Admixtures Association, Cement and Concrete Association, Cement Makers' Federation, Clay Roofing Tile Council, Copper Development Association, Lead Development Association, Mortar Producers' Association, Natural Slate Quarries Association, Sand and Gravel Association, Stone Federation and Zinc Development Association.
Manufacturers' catalogues must be a final source.

BRICK AND BLOCK WALLS

British Standards Institution

BS3921: 1985. Clay bricks and blocks.
BS4729: 1971. Shapes and dimensions of special bricks.
BS187: 1978. Calcium silicate bricks.
BS6073: 1981. Precast concrete masonry units.
CP121: Part 1:1973. Brick and block masonry.

Other sources

Publications and advice are available from various trade associations including Aggregate Concrete Block Association, British Precast Concrete Association, Calcium Silicate Brick Association, and Concrete Brick Manufactured Association. Manufacturers' catalogues.

IN SITU CONCRETE WALLS

British Standards Institution

BS5328: 1981. Ready-mixed concrete.
BS1881: 1970. Methods of testing concrete.
BS6100: 1984. Glossary of terms for concrete and reinforced concrete.
BS5328: 1981. Methods of specifying concrete.
BS4482: 1985. Hard drawn mild steel wire for reinforcing concrete.
BS5896: 1980. High tensile steel wire for reinforcing concrete.
BS4483: 1969. Steel fabric for reinforcing concrete.
and other appropriate standards listed in earlier sections.

RENDERED WALLS

British Standards Institution

BS5262: 1976. Code of practice for external rendered finishes.
BS1199: 1976. Sands for external renderings.
BS4764: 1986. Powder cement paints.
BS6213: 1982. Guide to selection of constructional sealants.
and other appropriate standards listed in previous sections and those defining specific sealant types.

Other sources

Anon. External finishes and rendering. Cement and Concrete Association.
Trade Associations. Various publications from British Adhesives and Sealants Association, British Decorators' Association, and Paint Research Association.
Manufacturers' technical literature.

STONE WALLS

British Standards Institution

BS5390: 1976. Code of practice for stone masonry.
BS6100: 1984. Glossary of terms for stone used in building and other appropriate standards already listed above.

Other sources

Trade Associations. Various publications from Natural Slate Quarries Association and the Stone Federation.
Quarrymens' technical literature.

TIMBER FENCING

British Standards Institution

BS565: 1972. Glossary of terms relating to timber and woodwork.
BS881/589: 1974. Nomenclature of commercial timbers.
BS4471: 1971/73. Dimensions for softwood.
BS5450: 1977. Sizes of hardwoods and methods of measurement.
BS4261: 1985. Glossary of terms relating to timber preservation.

BS6150: 1982. Code of practice for painting of buildings.
BS144/3051: 1972/3. Coal tar creosote for the preservation of timber.
BS913: 1973. Wood preservation by means of pressure creosoting.
BS1282: 1975. Guide to the choice, use and application of wood preservatives.
BS3452/3453: 1962. Waterborne wood preservatives and their application.
BS5056: 1979. Copper naphthenate wood preservatives.
BS5705: 1979/80. Solutions of wood preservatives in organic solvents.
BS1202: 1974. Nails.
BS1201: 1963. Wood screws.
BS2015: 1985. Glossary of paint terms.
BS1336: 1971. Knotting.
PA532: 1980. A paint system comprising an undercoat and gloss finish.
BS2521/2523: 1966. Lead-based priming paints.
BS4756: 1971. Aluminium priming paints for woodwork.
BS5082: 1974. Water-thinned priming paints for wood.
BS5358: 1976. Low-lead solvent-thinned priming paint for wood.
BS1722: 1972/86. Fences.
BS4102: 1986. Steel wire for fences.

Other sources

Trade Association. Various publications from British Decorators' Association, British Wood Preservation Association, National Federation of Painting and Decorating Contractors, Paint Research Association, Paintmakers' Association of Great Britain and Timber Research and Development Association. Manufacturers' technical literature.

METAL FENCES

British Standards Institution

BS1722: 1972/86. Fences.
BS4102: 1986. Steel wire for fences.
BS4: Part 1:1980. Hot-rolled steel sections.
BS4848: Part 2:1975. Hot-rolled hollow steel sections.
BS6323: 1982. Seamless and welded steel tubes.
BS1449: 1983. Steel plate, sheet and strip.
BS481: Part 2:1972. High tensile steel wire mesh.
BS6722: 1986. Recommendations for metal wire.
BS1052: 1980. Mild steel wire.
BS1554: 1986. Stainless steel round wire.
BS4483: 1985. Steel fabric.
BS405: 1987. Expanded metal (steel).
BS1485: 1983. Galvanized wire netting.
BS1474: 1988. Wrought aluminium bars, tubes and sections.
BS729: 1986. Hot-dip galvanized coatings on iron and steel.
BS1615: 1987. Anodic oxidation coatings on aluminium.
BS3987: 1974. Anodic oxide coatings on aluminium for external applications.
BS4921: 1988. Sheradized coatings on iron and steel.
BS4842: 1984. Storing organic finishes on aluminium.
BS3830: 1973. Vitreous enamelled steel.
and the standards for paints listed under Timber fencing above.

Other sources

Various publications are available from Aluminium Coatings Association, Aluminium Extruders Association, Aluminium Federation Ltd, British

Anodising Association, British Steel Corporation, Crafts Council, Fencing Contractors' Association and Society of Chain Link Fencing Manufacturers. Manufacturers' technical literature.

OTHER FENCING MATERIALS

Manufacturers' technical literature should be consulted.

GATES

British Standards Institution

BS3470: 1975. Field gates and posts.
BS4092: 1966. Domestic front entrance gates.
BS5707: 1979. Stiles, bridle gates and kissing gates.
BS1227: Part 1A:1967. Hinges for general building purposes.

Other sources

Publications from Guild of Architectural Ironmongers and from manufacturers.

CHANGES OF LEVEL (RAMPS)

British Standards Institution

*BS 5810:*Access for the disabled to buildings.
BSI Education Information, *Aid for the disabled*.

Aggregate Concrete Block Association ACBA
60 Charles Street
Leicester LE1 1FB
Tel: Leicester (0533) 536161

Aluminium Coaters Association
c/o British Standards Institution
Quality Assurance Section
PO Box 375
Milton Keynes
Bucks MK14 6LO
Tel: Milton Keynes (0908) 315555

Aluminium Extruders Association
Broadway House
Calthorpe Road
Birmingham B15 1TN
Tel: (021) 455 0311
Telex: BIRCOM-G-338024ALFED

Aluminium Federation
Alfed
Broadway House
Calthorpe Road
Birmingham B15 1TN
Tel: (021) 455 0311
Telex: BIRCOM-G-338024ALFED

Asphalt and Coated Macadam Association
see British Aggregate Construction Materials Industries

The Asphalt Institute
Asphalt Institute Building
College Park
Maryland
USA

Brick Development Association
BDA
Woodside House
Winkfield
Windsor
Berks SL4 2OX
Tel: Winkfield Row (0344) 885651

British Adhesives and Sealants Association
BASA
Secretary
33 Fellowes Way
Stevenage
Herts SG2 8BW
Tel: Stevenage (0438) 358514

British Aggregate Construction Materials Industries
BACMI
156 Buckingham Palace Road
London SW1W 9TR
Tel: (01) 730 8194
Fax: (01) 730 4355

British Cement Association
BCA
Wexham Springs
Slough
Berks SL3 6PL
Tel: Fulmer (02816) 2727
Telex: 848352
Fax: (02816) 2251/3727

British Decorators Association
6 Haywra Street
Harrogate
N Yorks HG1 5BL
Tel: Harrogate (0423) 67292/3

British Precast Concrete Federation Ltd
BPCF
60 Charles Street
Leicester LE1 1FB
Tel: Leicester (0533) 536161
Fax: (0533) 51468

British Standards Institution
BSI

Head Office:
2 Park Street
London W1A 2BS
Tel: (01) 629 9000
Telex: 266933
Fax: (Group 2/3) (01) 629 0506

British Steel Corporation, Research Services
BSC
BSC Swinden Laboratories
Moorgate
Rotherham
S Yorks S60 3AR
Tel: Rotherham (0709) 820166
Telex: 547279

British Wood Preserving Association
BWPA
Premier House
150 Southampton Row
London WC1B 5AL
Tel: (01) 837 8217

Calcium Silicate Brick Association
CSBA
24 Fearnley Road
Welwyn Garden City
Herts AL8 6HW
Tel: Welwyn Garden (07073) 24538

Cement Admixtures Association (CAA)
2A High Street
Hythe
Southampton
Hants SO4 6YW
Tel: Hythe (0703) 842765

Cement and Concrete Association
see British Cement Association

Clay Roofing Tile Council
CRTC
Federation House
Station Road
Stoke-on-Trent
Staffs ST4 2SA
Tel: Stoke-on-Trent (0782) 747256
Telex: 367446
Fax: (0782) 744102

Concrete Block Paving Association
Interpave
60 Charles Street
Leicester LE1 1FB
Tel: Leicester (0533) 536161
Fax: (0533) 514568

Concrete Brick Manufacturer's Association
CBMA
60 Charles Street
Leicester LE1 1FB
Tel: Leicester (0533) 536161
Fax: (0533) 514568

Copper Development Association
CDA
Orchard House
Mutton Lane
Potters Bar
Herts EN6 3AP
Tel: Potters Bar (0707) 50711
Telex: 265451 MONREF (quote 72:MAG 30836)

Crafts Council
12 Waterloo Place
London SW1Y 4AU
Tel: (01) 930 4811

Decorative Paving and Walling Association
60 Charles Street
Leicester LE1 1FB
Tel: (0533) 536161
Fax: (0533) 514568

Fencing Contractors Association
FCA
St John's House
23 St John's Road
Watford
Herts WD1 1PY
Tel: Watford (0923) 248895

Guild of Architectural Ironmongers
GAI
8 Stepney Green
London E1 3JU
Tel: (01) 790 3431
Telex: 94012229 GAII G
Fax: (01) 790 8517

Her Majesty's Stationery Office
HMSO
St Crispins
Duke Street
Norwich
Norfolk NR3 1PD
Tel: Norwich (0603) 22211
Telex: 97301
Fax: (0603) 695582

Lead Development Association
LDA
34 Berkeley Square
London W1X 6AJ
Tel: (01) 499 8422
Telex: 261286

The Mastic Asphalt Council and Employers Federation
Construction House
Paddockhall Road
Haywards Heath
RH16 1HE

Mortar Producers Association Ltd
MPA
Holly House
74 Holly Walk
Leamington Spa
Warwicks CV32 4JD
Tel: Leamington Spa (0926) 38611

National Federation of Clay
Industries Ltd
NFCI
Federation House
Station Road
Stoke on Trent
Staffs ST4 2TJ
Tel: Stoke on Trent (0782) 416256
Telex: 367446
Fax: (0782) 744102

National Federation of Painting and
Decorating Contractors
82 New Cavendish Street
London W1M 8AD
Tel: (01) 580 5588

National Paving and Kerb
Association
NPKA
60 Charles Street
Leicester LE1 1FB
Tel: Leicester (0533) 536161
Fax: (0533) 5¹4568

Natural Slate Quarries Association
Bryn
Llanllechid
Bangor
Gwynedd LL57 3LG
Tel: Bethesda (0248) 600476

Paintmakers Association of Great
Britain
PA
Alembic House
93 Albert Embankment
London SE1 7TY
Tel: (01) 582 1185

Paint Research Association
8 Waldegrave Road
Teddington
Middlesex TW11 8LD
Tel: (01) 977 4427
Telex: 928720
Fax: (01) 943 4705

Refined Bitumen Association
165 Queen Victoria Street
London
EC4V 4DD

Sand and Gravel Association Ltd
SAGA
1 Bramber Court
2 Bramber Road
London W14 9PB
Tel: (01) 381 1443

Society of Chain Link Fencing
Manufacturers
16 Montcrieffe Road
Sheffield S7 1HR
Tel: (0742) 500350

Stone Federation
82 New Cavendish Street
London W1M 8AD
Tel: (01) 580 5588
Telex: 265763

Timber Research and Development
Association
TRADA
Stocking Lane
Hughenden Valley
High Wycombe
Bucks HP14 4ND
Tel: Naphill (024024) 2771/3091/3956
Telex: 83292 TRADA G
Prestel: 3511615
Fax: (024024) 5487

Zinc Development Association
34 Berkeley Square
London W1